CU01551832

Technical Writing – Technical Comm

Author: Alicia Crowder

Contributing Authors: Dayne Hunt, Mike Bealle, Sherine Davis

Editors: Stacy Mandell, Scott Purde

Illustrator: Nick Chagouris

Cover art by Alicia Crowder

Copyright 2013

1st Edition

ISBN-13: 978-1502985378

ISBN-10: 1502985373

Figure 1 – Artist: Nick Chagouris – Copyright 2014

Contents

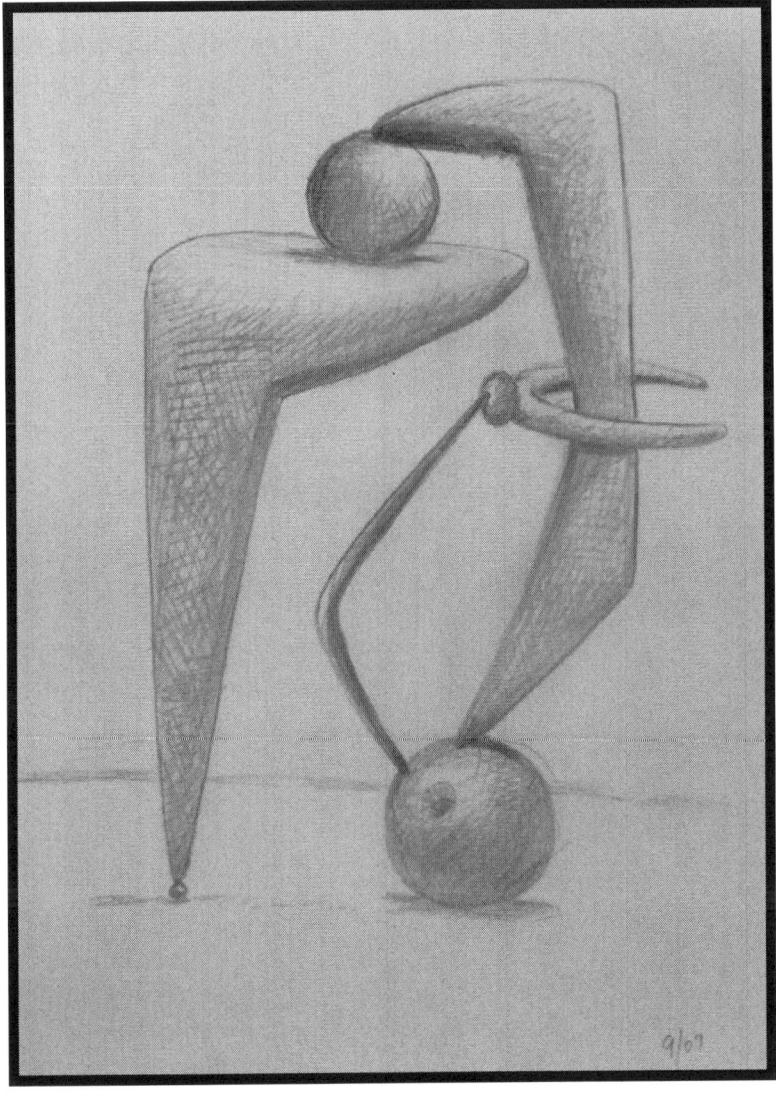

Introduction

This book has been developed to teach individuals interested in working as technical writers, and those whose job will include technical communication, many details of the technical writing profession.

Whether a technician or engineer whose career depends on the ability to communicate about, prove, and document aspects of work in written form, or a full-time professional devoted exclusively to the field of technical writing; information and instruction to improve technical communication capabilities are included within the pages of this book.

Real world examples of environments, applications, and challenges technical writers face on a daily basis have also been included. The purpose is to better prepare the new technical writer to jump into the field of technical communications with realistic expectations and knowledge as to how to provide the service of technical writing.

The reader of this book should discover what they may encounter while working independently and as a team member, how to handle some situations they will face, where to find the resources necessary to perform the work of a technical writer, and the specifics of how to write professionally in the technical communication field, as opposed to the work and rules involved in creative composition development, academic, or other genres of writing.

A focused student of this book will be equipped with the tools needed to understand how to approach technical writing, what details to be mindful of as they pursue this craft, what pitfalls and assumptions to avoid, and ways to have a better chance of successfully enjoying a long-term career as a technical writer by building and protecting their reputation and work.

The field of technical communication includes (but may not entirely be limited to) the tasks associated with technical writing. However, technical writers are sometimes also labeled as technical communicators.

The purpose of technical communication is to clearly and concisely provide information to the reader or recipient of the technical communication to aid in the performance of some task or reaching a decision or goal.

A technical communicator may work individually, or in partnership with other professionals of varying kinds such as sales professionals, marketing groups, operations professionals, designers, quality assurance professionals, engineering professionals, scientists of all sorts including medicine, software development firms, computing and hardware development, repair and maintenance groups, or other individuals in industry.

Becoming proficient at technical writing involves many of the same principles as other types/styles of writing. However, there are also many things about technical writing that differ from other types of writing such as creative, sales, or fiction writing. Academic writing is also very different as the readers of a technical document do not have the same focus as the readers of academic writing. Non-fiction writing is sometimes called technical writing, but not all non-fiction is necessarily technical writing.

Depending on your industry focus, your perception of technical writing may differ. If you are a computing technology professional, you generally will think first of writing related to computers and applications as the definition of technical writing. Separately, as a process, electrical, mechanical, or other type of engineer, you will likely think of technical writing in terms of your field of expertise.

As a standard, the rules of technical writing and technical communication apply to 'all of the above' regardless of whether you are looking at writing for the manufacturing industry, the oil and gas industry/energy industry, or the computing and/or software industry, etc.

The distinctions lie heavily in considering terminology, conventions, document type, and industry standards. Within each industry technical writing may involve more of a step-by-step approach, an instructional method, or may just be informative and written in a different style.

Often you will be gathering the information provided by engineers, designers, scientists, and developers. Your tasks may be to transform the complexities presented into clear, concise, much easier to understand written form. This makes you in essence a technical translator.

Translating technical data into readable, absorbable content, more accessible for a wider audience, requires developing a special skill set in terms of a way of thinking, processing information, and analyzing information you hear, see, and read to discover clarity in the details.

There are many unique aspects to the way the technical writer must think and in the deliverable the technical writer should submit as the end result of their efforts. The concept of consistency is one example of an interesting difference to note in technical writing verses creative writing or perhaps even other types of correspondence, journaling, and some non-fiction.

You may believe consistency is important in all forms of writing and you would be correct. However, there is more than one way to ensure your document is consistent. When learning how to write and when practicing creative writing, it is common to search for varied terms that mean the same thing.

Do not bore your readers by using the same phrases or words over and over again. If you are attempting to use repetition for emphasis, you can do this by finding several different ways to say the same thing. This is often effective in driving home a point, idea, or concept, while not making the reader feel as if they are trapped in a redundant loop. Additionally, people may need to 'hear' something a different way in order to fully grasp it.

This method can also be applicable in instructional writing; however, you need to be more consistent with your word usage when you are writing technical documents. It is important to prevent confusion in one who is attempting to learn.

Altering the terminology used will cause them to wonder whether or not you are referring to a separate concept or item than was previously mentioned in the text. This does not mean it is always inappropriate to provide more than one way to understand the same concept; merely that you must be careful with your wording.

For example, let a hammer be a hammer every time you reference it and do not refer to it as a gavel, mallet, or in any other way if you are still describing the same tool. A gavel is different from a hammer and a mallet is different from a gavel. You may have interchanged the words in your personal life, but be careful not to do so in your technical writing. Even if there is no argument against the word you are using having two different names; pick one and stick with it.

There are a variety of ways consistency is extremely important in all styles of writing. The difference pertaining to technical writing is the way in which you should utilize consistency.

The important thing to realize here is that if you are a very talented creative writer, you still need to advance your education in some respects before you jump into the field of technical writing. This will aid you in avoiding embarrassment and in producing quality documentation that accomplishes its purpose.

Technical writing involves conceptual differences a creative writer is not usually accustomed to considering. For that reason I listed the brief, but poignant example of consistency in the paragraphs above as part of the introduction to these technical writing lessons.

In the sections to come, you will be exposed to information related to the core competencies of technical writing. If interested in performing technical writing on a professional level, your study of this text will provide you with the knowledge to successfully act on that desire.

The effort does not end at the outset though. You must continuously work in a detail-oriented fashion and double-check yourself to ensure you haven't fallen back into old habits throughout the course of writing whichever type of technical document you tackle.

The information you will find in this book will include topics such as:

- The mindset of a successful writer
- The writing preparation process
- Research for technical writing
- Grammar and writing styles
- Technical writing standards and conventions
- Formatting technical documentation
- Indexes and references in technical writing
- Editing technical documentation

In the research section of this book, you will find a list of many specific sources to which you can refer to gather information for a variety of industries. At the end of the year 2014 these are all current resources.

Be aware that while you will be equipped via the purchase and reading of this book with a long list of detailed resources to aid you in your research, there is no way I could have reasonably captured every single research resource available. Rather, I have merely included the specifics of where and how to obtain information by randomly including a sampling of credible sources you can use for your work.

You'll read this later in the book, but I stress heavily now as well the importance of having someone else review your work. If you are new to technical writing, this may make the primary difference between succeeding and failing (*failing of course only if you give up and quit*) to become a skilled, proficient technical writer who is sought out (*offered work now and in the future*) in your field of expertise.

Even the most seasoned writers benefit greatly from receiving editing assistance from professional editors in their industry, as well as at times from friends and family.

Caveat: Don't simply accept all changes blindly after passing your technical document to anyone qualified or unqualified to edit technical writing. Review their changes, use discernment, and refer back to what you've learned in this book and in other technical writing instructional materials and writing style guidelines before accepting each change.

Ensure you are not lead astray by uninformed, good intentions. Even extremely experienced writers have blind spots and are prone to arrogance about their level of knowledge such that they may believe something to be the 'right' way, when in fact, it is not. It's possible they remember incorrectly, or that their education is out of date, or their field of expertise may be separate and involve different rules.

For example, a professional journalist may be capable of providing exceptional feedback, but is generally not qualified to provide industry-specific technical editing services if their experience is limited to newspaper writing. Whatever your exposure, walk that line of being open to even the most intensive critique, while using your best judgment along the way.

Remember, the buck falls on you when you turn in or publish your document. Therefore, the level of accuracy and style of writing needs to be such that you are comfortable with it being part of the written record of your existence and career.

In a personal sense I consider myself an artist who has an offbeat sense of humor; a desire to change the 'norm' of acceptable standards, and a blatant disregard for the opinions of others who would tell me what should and should not be published.

As a result, when it comes to my creative writing and purely for my own satisfaction or goal achievement, I've published text some would never have allowed to see the light of day. I'm not generally ashamed of this, nor would I reverse the clock if I could go back and do things differently.

From my indie artist perspective it was a positive growing experience and an opening myself up to critique, developing a sense of expression, testing out style variances, and I even consider it a demonstration of courage. In some perspectives it may have been laziness or disregard, but considering the subjective nature of art I leave it to be whatever it is and was.

However, in the professional world where I have written for hundreds of other companies in technical writing and many forms of copywriting, I always check my subversive behaviors at the door. Those who rely on me to help them increase their business income and the people who depend on me to provide helpful, instructional materials they can comfortably share with all in their industry, are not disappointed in the results they achieve as a result of hiring me.

The rules for different types of writing are not identical. Further, the rules of writing for different companies may distinctly change. For many written efforts outside the realm of technical writing, feel free to make up your own rules. That is an opinion I've developed based on my study of published, popular, and not so popular literature of historical and modern times.

You are the only one that has to deal with the consequences of your choice in a truly personal way. But it's time to step up and conform when you are providing technical writing services for other businesses. You certainly don't have to conform to rules that betray your sense of quality writing. However, be prepared that if you want to avoid certain types/styles of writing, or do not want to conform to the desires of your client or employer, you may be in essence turning down current and future work opportunities.

The standards and guidelines exist because they are necessary or desired by your employer. If your employer or editor desires these guidelines be followed, that in and of itself, makes them necessary. When these change, you need to adjust with them. It's as simple as that. I have had to place commas in areas I did not believe they belonged and change wording styles to match what I considered ignorant perspectives in the past. I have also had to bounce back and forth between contrary ways with regard to the same document because of the differing opinions of editors within the same project. This is part of life as a professional writer. It can be annoying but what is most important is to remain patient and if you have to put additional, unnecessary time into a project because of back-and-forth changes make sure you get paid for your extra time.

If you are not working on an hourly basis, set a limit on free re-writes and edits and make sure your client understands this at the outset. Your contract or written agreement should clearly note this point. This will improve the chances of your client behaving more efficiently and sets the foundation for a better long-term working relationship.

And one last note I must include in this introductory section relates to the word 'creative' as I've been using it and will continue to use it throughout this book.

There are many different definitions of the concept of creativity. It is true that in some ways you can be creative in technical writing. However, you should not attempt to do this with a casual writing style, nor a flowery and overly descriptive writing style.

When defining the dimensions of a table as part of an assembly instruction sheet, your reader does not need to know you feel the apple red color of shiny latex paint on the dark walnut wood of the table set under the shadow of the pillars creates a haunting glow. That would be a distraction and including those details in the text may at least double the amount of time it takes your readers to complete their assembly project. Save all of that type of creativity for your personal masterpieces or for clients who want that kind of thing and leave it at the door when you are writing technical documentation.

There are of course exceptions to every rule and some might consider those 'idiots guide' books and those 'head first' books to be technical writing of a sort.

They would not be incorrect in calling those books technical. However, unless you are filling that kind of a niche, or are just writing a technical document purely for yourself and other members of the public who are buying your book directly or reading your articles and so on, then you need to fall in line with the standards of technical writing.

You can certainly try to convince an oil and gas company that their project design specifications should be more entertaining and colorful to make life more 'fun' for all of the engineers and project managers who need these as a reference. But don't say I didn't warn you if you find yourself out of the running for the next technical writing job you could otherwise have landed thereafter; or if the publisher you are chasing rejects your document with an impersonal form response.

Many have had to conform and follow rules they did not agree with throughout their life. As a result, they often don't have an interest in bothering with the elitist attitudes we artists can project. (I use the term 'elitist attitude' not so much in a way you would understand if you look it up in the dictionary, but more in a slang way commonly thought of, which is an attitude that you are somehow above following the same rules everyone else has to follow.)

Many have survived and sometimes succeeded in big business by putting their personal desires aside and conforming to the extreme. This may cause them to be less sympathetic to the cause of those who attempt to rebel in business. What is cute to you may easily be worse than an annoyance, but actually a hindrance to others. This will not only reflect badly on you, but also on your employer who contracted or hired you and potentially on the organization you are writing for as a whole if one of your subversive efforts makes it past the final edit and review.

And with that little speech, the introduction is concluded and you can dig into the meat of learning the lessons of technical writing and technical communication.

TIPS:

- **Professional technical writing involves following rules.**

- **Avoid flowery writing in technical documentation.**

- **Words like 'holistic' should at the least be reconsidered in technical writing and never used frequently, at best never used at all.**

- **Overuse of a word in common/pop culture business conversation dilutes the effectiveness of the term.**

- **Make sure you are paid for your time.**

The mindset of a successful writer

Figure 2 - Artist: Nick Chagouris - Copyright 2014

Awareness is a key component in becoming a successful writer. As will be discussed in more detail later within this book, one important aspect of the mindset you'll need to have is a reader focus. You will likely be writing for readers with varied educational, cultural, and technical backgrounds, as well as for readers within organizations that each have different roles and job functions to consider.

Additionally, it is important to recognize that whether you are a full-time technical writer, or are in a technical trade of some sort requiring you to document the work you've accomplished or have been associated with, how you write and what you write is an indication of how well you have done your job. Your abilities, for example, as an engineer, are in part judged by your technical writing skill.

In order to understand why you would consider mindset in order to be a successful writer, a few things should be explained. By 'successful' in terms of the general concept of 'being a writer' may mean someone who is a prolific writer. Doing a lot of writing increases your opportunities to become published by others, or even just having anything available to self-publish. If you want to consider yourself a successful writer because you wrote a poem once, that's your call.

However, there are multitudes of people who have sacrificed a great many hours of sleep in order to turn out a large amount of writing for publication and personal fulfillment throughout the years. It takes dedication and hard work to join the ranks.

Those who have earned the title are no better than you. They have simply worked harder than you up to this point. If you work as hard as they do at the craft of writing, then you have the right to call yourself a successful writer based upon any definition you choose.

If you haven't spent more than 10 hours of your life writing; if you haven't had to struggle with your internal editor; if you haven't had to deal with criticism of others who've never tried to write anywhere nearly as much as you, if at all; if you've only ever written school assignments; then please feel free to keep writing and to keep the dream alive; to do as much or as little writing as you wish. But it may be a good idea to wait a little while and put some effort into your craft of writing before claiming the label of writer. The point is to leave your arrogance aside and enter the working mindset; carefully crafting and refining your skills until one day there can be no doubt you have earned your title.

Some consider the term 'success' to be in direct correlation to money. That's fine. Most of us wish to be that kind of successful writer. However, the true success is in continuing day after day, year after year, despite acceptance, rejection, thousands of Internet page views, or only two.

The initial successful feeling related to finances comes when people reach out to you because they've enjoyed your writing and want to pay you to do more.

Alternatively, even without the monetary benefit, the true successful feeling comes when others let you know they've enjoyed your writing and it in some way impacted or improved their day or life. However, this may never happen in terms of technical writing, so don't base too much on this aspect of the experience if you are merely forced to document your technical work.

The more personal feeling of success comes when you've felt the pain of forcing yourself to push through a difficult writing assignment and have reached the other side despite the voice in your head that urges you to put down the pen and paper, or step away from the computer and do something else that will take less time, energy, and seems less intimidating.

I'm not going to sugar coat this. Every time you put that laptop on your knees and start to type; every time you pick up that pen and paper; you are at risk of spending minutes, hours, or even days working diligently through a document or manuscript you complete and experience a momentarily deep sense of satisfaction completing, only to re-read it later and believe that what you wrote, or the way you wrote it, is total garbage.

The up side to this though is that once you have pulled those words out of your head and put them onto the written or digital page, you have accomplished the first step and provided yourself with a foundation on which to improve. It is not uncommon for a few edits and re-writes to absolutely transform a page full of garbage into a masterpiece.

In the professional, business world of technical writing, to be 'successful' as a writer is more heavily based on your ability to effectively convey the necessary information required and/or desired by your readers. This means writing something easy to understand by the appropriate audience, that provides the information promised and/or expected, and that does not create distractions via dramatically improper formatting and/or spelling and grammatical errors.

When you write a fictional piece or a technical document, you need to adopt the mindset of a successful writer. Even if you have only written tweets in your life previously, or you have actually composed multiple anthologies, it makes no difference in terms of the need to bring yourself into the right mindset.

This means spending time writing and not allowing yourself to become distracted until you are at an actual stopping point. Also, there is the issue of the internal editor you will need to manage in order to successfully write.

Each moment of completing a writing assignment someone else gave you or you gave yourself is a success. You can certainly say, "Today, I was a successful writer!" every time you begin and complete a writing task.

Distractions are a funny thing for a writer. In some instances, distractions can be the downfall of a would-be writer, or a writer who has been successful, yet fell into a period within which they abandoned their craft for days, weeks, months, or even years. However, distractions can also sometimes help you break through a stale moment in your mind.

At times distractions can also be just the thing you needed to awaken those syntactical functions inside of your brain that will result in words spilling out onto the page. Distractions in life can sometimes generate the stories you will tell and spark the idea that gets the job done.

But even when distractions have a positive side to them, you must at some point set them aside and avoid them like the plague. Stop all else and write what it is you must write at this moment. If you attempt to start 10 different stories at once you will have 10 different unfinished stories until you go back and re-work them one at a time.

I know this because it is the story of a large percentage

of my last 30 years writing. (Yes. I am older than 30...) I resume the work of each eventually, but maintaining focus on one story or document at a time is a more efficient way to complete each project. If necessary, jot down a few notes when other ideas arise and then return to the immediate task at hand.

If you are working on a technical document, as much as possible, avoid distractions unrelated to the topic on the table. Sit down to write and push out all distractions entirely.

A Buddhist monk once told me to visualize myself sweeping with a broom in my mind. Let unrelated thoughts flow in and flow back out. If they linger, sweep them away. They are not lost forever.

If you think you might benefit from a few distractions, wait to open yourself up to them until after you have at least finished your first draft.

True technical writing jobs, whether these are in-office

jobs, telecommuting jobs, or freelance jobs, all have deadlines. If you allow distractions to create a foggy sense of writer's block, you lose money and potentially respect and future opportunities.

An appropriate mindset allowing you to establish yourself as a successful technical writer involves a focus on quality. You must ask yourself questions such as: Does the document I'm creating enable the reader to do something? Does it enable my reader to take an appropriate action? Does the document I'm creating convey authority? Is it thorough? Is it accurate and honest?

What is the first impression my technical document will provide? Does it appear to be neat, organized, and readable? Is the necessary information accessible to a target audience such that they are able to get what they need from reading the section pertaining to them? Do they have a clear indication as to how to locate the section they need to view?

Is the purpose of my document clear? Does my writing make sense by being in the correct order for the reader? Bottom line: Is the document I'm writing usable?

Set yourself up with a project management mindset.

In project management there are stages aka phases. Each stage has a specific set of tasks. Each set of tasks accumulate in groups as milestones to reach within the context of a schedule.

If you've never received a project management education of any kind, I strongly urge you to do some searching on what is involved in project management and perhaps even take a project management course. It will definitely help you in all aspects of your business life and advancement, including your career as a writer if you are exclusively focusing on that.

TIPS:

- **Use a project management focus and break your work tasks into stages/phases.**

- **Put distractions in their place.**

- **Keep writing!**

Communicate and be firm with your internal editor

Figure 3 – Artist: Nick Chagouris - Copyright 2014

Your internal editor may be your best friend and/or your worst enemy. When you are developing the mindset of a writer, you need to treat your internal editor like a family member. Listen to what your internal editor has to say. Acknowledge the legitimate points as just that.

Next, let your internal editor know you cannot answer the phone for the next few hours because you have to focus on your work project, but assure your internal editor that you will return its call as soon as this stage is complete.

In communicating this way, you are better able to relieve yourself of the constant nagging, or as with the example, the ringing of the telephone from the family member/internal editor who wants to know if you are okay and will call them back as soon as you are able.

Turn off the volume on the phone in case your internal editor or family member is feeling a bit neurotic today and doesn't want to be patient.

You must take charge of your own mind and give yourself permission to make mistakes. If you do not do this, you may never complete your first draft and therefore will not be able to reach your scheduled milestones.

Set a project phase specifically for imperfect writing. This may help you to avoid the editing interruptions until the appropriate time. Don't let your internal editor convince you it needs to hold your hand throughout this stage of the project. It will squeeze your hand way too hard and you'll exhaust yourself in the process of attempting to follow its precise lead. Don't even allow it to have a loose grip. Shut it down and lock it out of the room! The micromanager must be banished for now.

Maintaining focus during consistent periods of writing time will assist you in becoming more comfortable with this process. In time, your stages will evolve and your natural instinct will improve.

Many of your first drafts will begin to look more like your second and third drafts of years past. This will not be the case every time, but it is a process that gets easier over time depending on the type of writing you are doing.

With technical writing there are specific formats you will become more comfortable using. This specific type of writing will become much more natural with increased experience. Additionally, your internal editor's volume will be lessened by educating yourself through the study of guides such as this one and others, as well as through practice of what you are learning.

A relationship of trust will be built with your internal editor. It knows you never forgot to call it back as promised in the past, so it will await your call with more ease than when you first began as a writer.

Writing is a lot of work and it is easy to become impatient. Calm yourself and tackle your tasks diligently.

No matter how far along you are in your career, you must force yourself to hold firm to your agreement now and forever that during the appropriate stage of writing, you are allowed to make (and temporarily leave) mistakes.

Do not go running to get the paper towels when the milk first spills on the floor. Leave it on the floor. You've put the protective coating down so that it won't damage the hardwood by scheduling the editing and rewriting stages into your project plan.

At this point I'm going to interject for just a moment with a note about being creative with technical writing in comparison with some of the illustrative writing I'm doing here in this book of technical writing lessons. You might have stopped and thought, "Wait...isn't the writer of this book doing exactly what she said not to do in technical writing?"

Yes, to a certain degree I am. However, while this is an instructional and informative book, it's not really what I would consider to be a technical writing assignment in terms of the standards described.

I have a little more leeway in how I can write within the sections of this book than I would have if I were writing for one of my clients. Therein you find the gray area I was referring to earlier. In this case, I get to do what I want as long as I provide you with a great deal of helpful information throughout the pages of this book.

You may be a busy professional, but you must slow yourself down a bit during the educational process to ensure the full concepts seep in enough that you will employ them in your work.

As long as I am presenting accurate information to help you reach your goal of becoming an effective technical writer, I am accomplishing this goal. This is not a case of 'do what I say and not what I do' so much as it is a case in point of the differences involved in separate types of writing.

TIPS:

- **Do not be controlled by your internal editor.**

- **Take heart! It will become easier with time.**

- **Practice may not make perfect, but it makes the process easier.**

The imperative question

As a technical writer, whether writing for the record or writing for a reader, always question the content of your work. You do not have to do this during the first draft stage, but it may save you a lot of time if you have the correct mindset from the beginning. This is not a contradiction to the point of silencing your internal editor at the appropriate time.

Keep one question in mind and ever present while you are editing your work. Is this information necessary? Always focus on providing what is needed to be effective and eliminating information not required for creating each specific technical document.

The concept of précis is paramount to delivering quality work as a technical writer, but your caveat is to avoid creating a doorway your internal editor may enter through, thereby blocking the flow of your writing.

When editing the technical document you've written,

look at it in sections first. Is every section included actually necessary to achieve what should be the singular purpose of this specific document?

For students of the programming language C# and others, SOLID is a principle of software engineering architecture. SOLID is an acronym that will not be discussed here beyond the reference to what the "S" in SOLID represents. "S" in SOLID stands for SRP, which is the acronym for Single Responsibility Principle.

SRP is a programming principle that states the importance of précis in code. For example, one class should not be responsible for more than one thing. This concept of separating responsibilities is a good example of how you should look at your technical document.

You may have to provide instruction and/or information on more than one thing. However, your document must be absolutely clear in order to be effective. In order to be clear, each section must be singularly responsible for communicating a specific message or instruction. If you find you have blurred the lines between where one type of information begins and one type ends, then you need to re-evaluate the formatting or layout of the content of your document.

Additionally, this careful single responsibility review provides an opportunity for you to analyze whether or not each bit of information you've included is actually associated with and necessary for meeting the requirements involved with the responsibility of the technical document as a whole.

Have you strayed off topic?

Have you overzealously shared more information than is needed for the busy professional who must scan through your technical document to glean the facts required for their job?

Is there any information included that would cloud the vision of the one seeking answers from that which you've written?

After determining which sections to keep and which to discard, then review the content within each section. Ask the same question once again. Does each paragraph cover the correct topic and contribute in an essential fashion to delivering the message of its section? How about each sentence within each paragraph?

Retaining the focus of only providing the truly needed information is a matter of mindset for the effective writer in delivering quality documentation useful to the intended audience and in turn, in delivering a reputation as worthy to contact for future work.

TIPS:

- **If it doesn't need to be in the text, don't include it!**

- **Keep each section focused and on topic.**

Information gathering filter

How do you know what to write about? If you don't know what to write about, the idea of becoming a successful writer may seem challenging or impossible. If you cannot generate ideas on your own; with enough study, searching, and practice, it is possible discovering these ideas may eventually become a natural part of who you are.

In some genres of writing, you will simply receive assignments telling you what to write about. Yet, you still need to be able to process thoughts around topic. Even if this comes naturally to you, practice is still required and foundational to the evolution of a talented writer.

Once you have adopted the mindset of a writer, it becomes a part of you and your daily interactions. As far as technical writing is concerned, this way of thinking is particularly important when you are in the midst of research and interviews.

You need to be able to pick out the pieces and the specific words you'll be including in your technical document. You must listen for the subjects deeper than the words used, inquire further and develop written material around those important topics and details that will give your document more meaning and make it much more useful to the readers who need that information for their own purposes.

Research and interview with directions and focal points in mind, but be careful not to be closed off to information you can obtain that you didn't yet realize you needed. You may very likely not be the (SME) subject matter expert. You are the messenger. So be careful not to miss out on fully gathering the message.

TIPS:

- **It can be helpful for a writer to have something of an analytical mindset in order to evaluate subjects and discussions, filtering out the pieces**

of focus for writing material.

- **Diligently seek out your SME's when possible and appropriate for the task at hand.**

So alone…

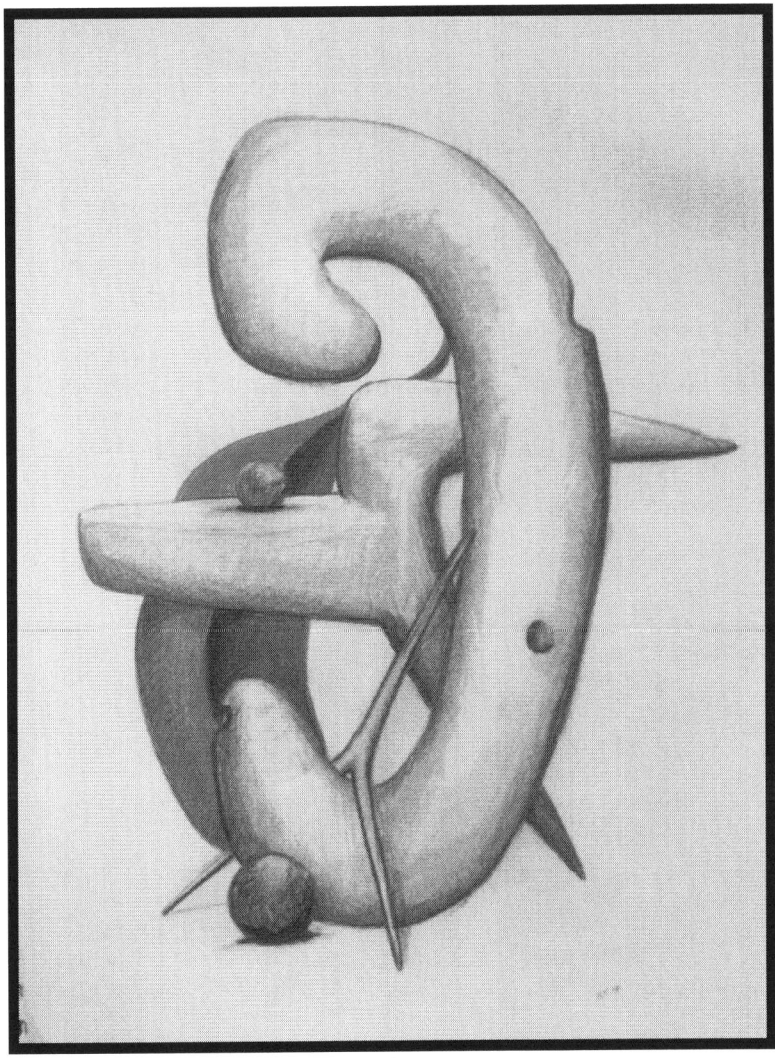

Figure 4 - Artist: Nick Chagouris - Copyright 2014

Technical writing is a solitary task for the most part. When performing technical writing, once the interview is over, if there is even an interview involved, which there may not be, and when you are not discussing requirements with the person who is paying you or a team involved in a project your writing is only a part of, you are back on your own again.

If stuck in the mindset that you require the constant involvement of others in order to get your job done, you may have trouble becoming a successful writer.

You have to be independently minded and comfortable spending hours on end without further interaction with others until you've completed your assignment. It would be advisable to close your email and turn off your telephone while in the middle of a writing session.

If you allow yourself to check emails, texts, and answer phone calls while in the middle of writing, you are going to take a great deal longer to complete your task and you might lose your train of focus such that what would have been a much better quality first draft now becomes disjointed and requires more extensive editing because you were not maintaining the best writer's mindset by recognizing your need to be alone and undistracted in order to write successfully.

TIPS:

- **Leave the phone in a different room or simply turn it off!**

- **Be independent and secure enough to spend hours alone if you want to be a technical writer.**

The need to prepare

You could just halfhazardly jump into your writing project. People do that all the time. But especially when beginning a technical writing project it is important to maintain a preparatory mindset.

Quality and attention to detail are of paramount importance when writing for the detail-oriented technical audience. Additionally, you must remember your reputation is your living.

Without a reputation for producing quality work and meeting deadlines you will have difficulty maintaining a job as a technical writer, or acquiring new clients as an independent technical writer.

With this concept of preparation in mind, we advance on to the next section of this book, which is all about preparation.

Writing preparation process

Part of the writing preparation process is considering who your audience will be and what their purpose is in reading the documentation you are developing. This means thinking about what information the readers are going to need in order to make the documentation an assistive tool for them in completing their associated objectives.

Additionally, why you are writing and what you want to achieve is as important as learning about your readers. These thoughts should all be parallel in your mind as you prepare to compose any document. In some technical writing circumstances your role within the organization you are writing for, if writing as an employee, should also be an affecting aspect of how you will phrase/word certain sentences and may be something you include when considering the task you are about to undertake during the preparation process. This can and should be reviewed during the editing portion of the work as well, but is good to keep in mind from the outset.

Determining the appropriate formatting for your technical document is also a part of the preparation process. You might want to include pictures as an aid and if so, it may be helpful to gather the images together before completing the text for your first draft. More information on formatting a technical document is provided in the formatting section of this book.

Full awareness of deadlines is also a critical element of the preparation stage. Determine what you have time for and how you will fit any interview time, formatting time, image gathering time, and of course, actual writing time, followed by editing time into the allotted slot you have between the present moment and the assigned deadline. This is part of the project management mindset and the creation of a schedule with stages/phases.

When planning to write a technical document, as well as when quoting a price for the creation of that technical document, you must acknowledge that each and every step involved takes time. If you aren't careful, you will end up feeling rushed and underpaid, and your feeling will be absolutely correct.

Preparing to write a technical document is something you need to look at from a variety of angles and make sure you do not undervalue what you are doing both for the sake of your own finances, as well as for the sake of your sanity and the actual quality of the work you are more likely to produce.

Organization, which will be addressed more as you progress in reading further in this book, is another aspect of your planning process. While planning it is good to also include the organizing steps. As an engineer, scientist, or other form of technician, this is also especially important to bear in mind. The planning process for any job should include gathering what you need for the writing portion of your work. If you are new to technical writing; studying to pick up a few tips to help you along your way is also part of the preparation and planning stage.

When creating your first draft, you may wish to create all of the main titles. Figuring out what these will be may be a part of your planning process. If you choose this method, completing your first draft can be an easier process because after that it will be more like a job of filling in the blanks.

TIPS:

- **Recognize there is value in what you do!**

- **Be reasonable so you can actually get work, but don't let yourself be underpaid.**

Project plan

One way to manage the writing preparation process is to develop a project plan for all steps involved in the development of your assigned technical document.

This first step of creating a project plan might seem like something you do not have time to do, but in the long run it will save you time and headaches. Your writing project can easily be a much more successful one if you have a path to follow such as that which you'll have outlined within your project plan.

Additionally, you'll be able to use this project plan as a template for future technical writing tasks.

With fiction writing, many instructors encourage their students to develop the characters as a step unto itself.

There is a similar task involved in technical writing. However the characters you need to write about are not fictional in this case. Also, you are not writing about them in a character development fashion. If you take a moment to craft an analysis of the readers who will be using your documentation, it will help you to provide a better experience, more attuned to the needs of your audience.

Staying on track with technical documentation requires dedicated focus on the task at hand. This is particularly true when writing technical documents spanning multiple pages, if not hundreds of pages.

Spend time actually writing down qualities and relational facts about your target audience and you'll end up with a much more effective product.

The basis for creating your project plan should be a review of a series of 'why' and 'how' and 'what' questions.

- What is this document for? (What is it going to be used for? Why does it need to exist?)

- What needs to be included in this document? (What kind of information am I gathering and writing about?)

- What is my deadline? (Do I have a soft deadline and a hard deadline, or just one big end-all be-all date that this has to be completed by?)

- How are any expenses involved in the creation of this documentation going to be covered? (Am I doing anything as part of the preparation of this document that will cost money, or is all the expense involved purely a matter of my time?)

- What are the expectations of my client? (Does the person paying me to write this document have any unique requirements outside of the norm for technical documentation? Do they have a different opinion about what a technical document should look like and is there an industry standard they are expecting me to match to any degree?)

Creating a project plan always involves the delineation of objectives. In order to assess the goals involved in achieving success in your technical writing project, knowing your audience is an essential key piece of knowledge to incorporate into your writing preparation process.

A different type of audience will mean a different type of technical document altogether.

This information is also a determining factor in the size of technical document you plan to write.

Craft your project plan with headers such as:

- Project Name
- Audience
- Resources

- Deadline
- Budget
- Format/Style
- Subject/Topic
- Contact Info
- Additional Notes

Laying out details such as these will help you stay on track after you've started your writing project and will help to ensure you are clear about your objectives. It will also provide you with a handy reference for future projects that come up.

You can always look back at former project plans to help spur additional ideas and in case you need to contact any of the same Subject Matter Experts (SME) again.

The process of developing a project plan definitely includes creating the list of people you will need to contact for information. This is a key point in determining the amount of time within which you can reasonably complete your tasks.

If you make sure to take care of these steps and the full development of the project plan up front, you'll have more ability to reshape the deadline without the embarrassment of later realizing you couldn't reach all of the people you needed to speak with on time, thereby losing the ability to complete your document within the allotted time frame. People are much more understanding about scheduling changes that occur weeks or months before they are expecting a deliverable, verses days or hours.

Additionally, you have to remember that you are not working in a vacuum. There are likely other people who have jobs to do that are reliant on the completion of your work. This means that if you don't let them know the date change in advance, they then schedule their workload based on incorrect dates and a domino effect occurs.

TIP:

- **Developing a project plan is an important part of being organized and creating high quality work.**

Readers | Get the details

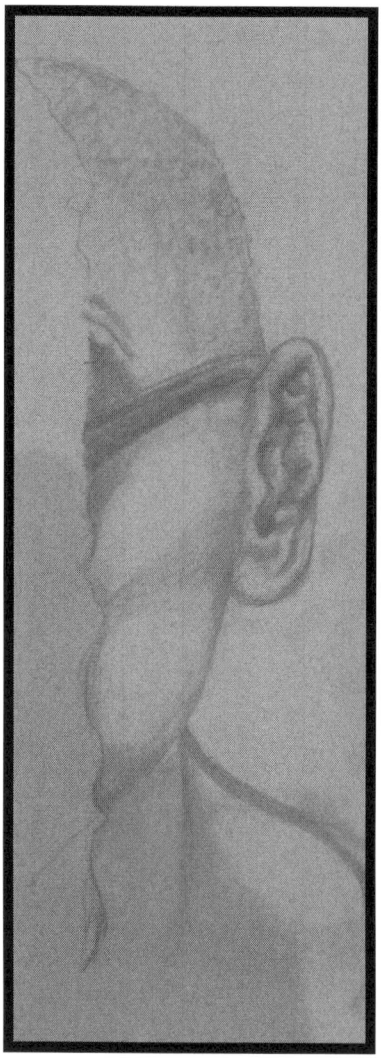

Figure 5 – Artist: Nick Chagouris - Copyright 2014

You don't necessarily need to know your reader's favorite foods, time of year, and color in order to craft a reader's analysis that will help you generate the best documentation to meet their needs.

The type of information to include should relate to their uses for the documentation and their prior knowledge of the topic. Consider the reader's frame of reference.

If you are writing to an audience that needs information from the ground up, you'll need to include that level of detail in the documentation you write and so will need to prepare to write in a way that teaches the reader what it is they are reading about and that progressively helps them in their transition from novice on the topic to more intermediate or advanced levels.

An audience of the skill level expected to already be familiar with all of the terminology and perhaps already heavily experienced in the related trade will need a document that is more supplemental, or perhaps one that highlights a facet of their trade, which may be entirely new to them, even though they are already familiar with all surrounding pieces of information and comfortable with the relative syntax.

In this case, you'll not need to, or want to, include the same level of detail or step-by-step instructions in connection with the more elementary aspects of the topic at hand. For some audiences, too much introductory information clouds the water and makes it more difficult for the reader to find what they are looking for in your document.

Unnecessary information can tire your reader and weaken the effect of the meaty part of your composition, perhaps even causing you to lose your audience such that they put down the document, book, computer, or e-reader and never pick it back up again before even seeing that you have included anything they would have found useful or informative. This too, can harm your reputation as a credible technical writer.

However, it is also very likely that you'll be writing for a mixed audience. The best solution in this case is to prepare to write the document in such a way that it alerts the readers to the sections which are most appropriate for them and provides the additional information to bring the less experienced readers up to speed in a separate section so as not to disrupt the flow for those with the educated background that purely need to acquire this information.

Determining these factors will assist you in choosing the formatting and in planning out the sections of your documentation that you will need to write within.

Getting to know your readers is called conducting a reader analysis. The reader analysis work is an action step that is part of what is called a work breakdown schedule.

TIPS:

- **Always get to know your readers!**

- **Never assume you know for a fact who your readers are. You might very well know, but double-check just in case.**

Work Breakdown Schedule

As mentioned in the section above, conducting an analysis of your readers is an action step in the work breakdown schedule. All of the pieces of the work breakdown schedule are action steps and the reader analysis is just the first of these.

The next thing you will want to do is review your list of subject matter experts and arrange times to meet with each of them. You can meet with them in whatever way you choose if they agree, but it is fairly common for these SME meetings to be conducted via telephone.

The step after the interviews should actually be started in a parallel fashion with the SME step because you will have gaps of time between interviews and this will help you in more quickly completing your project. You certainly can wait until your SME step is completed, but it is important to make sure there is a lot more lead time before you have to turn in your first draft, or finished writing project in order to separate these steps out sequentially in that fashion. There will often be a lot of downtime as part of that process.

If you are managing several projects at once though, separating your schedule out in this kind of sequential fashion might be quite viable and enable you to better use that downtime on whatever action step you have reached in a separate project.

Designing/formatting the document is this next action step referred to above. Also in this step you'll want to go ahead and choose which graphics, if any, you'll be including in the document. The decision to include images in your technical document is dependent on a number of factors.

The primary differentiating factor is directly associated with your role as a technical writer. If you are part of a team and are turning the technical document in to another person on your team for them to do their piece, you might not be assigned the task of including graphics in your technical document.

In this instance it would be more like stepping on someone else's toes if you went out and spent time picking out pictures to include in your writing piece. It all depends on the context within which you are creating your technical document. If you are merely a writer within a team then just do what you do best, write and edit and leave the rest to the graphic designers or other roles involved.

Now speaking of writing…That is your next action step! Write your first draft using the guides you created for yourself in your project plan and with the content you've derived from the discussions with subject matter experts and any other research you may have put into the preparation process.

After you've completed your first draft, it's time to be smart and hand it over to someone else for review. Don't trust your own judgment entirely by simply writing and editing without any other potentially more objective eyes scanning the document.

Do not forget to use spell check before you turn over the document so that you aren't wasting the reviewer's time with petty matters a computer program could easily have fixed.

Even typos and grammar issues that are not caught by the spell check application can easily be missed by the individual who actually created the first draft simply because they have a very clear picture in their mind of what the page is supposed to say, so much so that their eyes might deceive them when reviewing their own material.

To you the page might clearly state everything the way it should be stated with perfect spelling and grammar while the reality could be much different.

However, the final action step prior to turning in your work is definitely to proofread and edit your document even after someone else has reviewed it and potentially even thoroughly edited and proofread it.

What is the difference between editing and proofreading? There are different types of editing to be sure, but the basic separation between proofreading and editing is that proofreading focuses on basics like grammar, whereas editing will potentially include checking to make sure the words are used correctly, looking at the flow of the document, and a list of many other pieces can be included in editing as part of working to make the document as excellent as possible.

For technical writing, editing should include things like tone, professionalism of wording, proper descriptions and instructions provided, and terminology to match the industry the document is created for, among other things.

TIPS:

- **Get someone else to help you proofread and edit your work whenever possible.**

- **Your first draft should not be your final deliverable.**

- **Do the work you are assigned to do and allow your teammates to do their jobs as well.**

Time Management | Technical Writing

The time spent in researching and interviewing for your technical writing project is going to be one of the biggest time taking pieces. A recommended allotment of time to schedule for the information gathering portion of your technical writing project is basically close to half of your schedule.

Depending on how diligently you plan on working on this one technical writing project and how many other things you intend on doing each day while you are working on your project will affect the amount of time you should allot to the actual writing process itself.

You can easily average about 2000 words a day, or about four pages per day dependent upon how many other tasks you are attempting to accomplish each of those days. Therefore, if you have determined up front how many pages your technical document should be, perhaps by determining approximately how many pages will need to be in each section, then you can successfully create a time plan for your writing.

Be sure to give yourself a little bit of cushion in any schedule you create because time and unforeseen occurrence befall us all.

Therefore, if you think it should take you two days to complete the writing portion, give yourself three or four days in your actual schedule. Just don't let that cushion you put in encourage you to procrastinate thinking you have more time. That's one of the top causes of not being able to complete a writing project on schedule.

Some professionals advise only allotting about five percent of the time in the schedule to editing and rewriting, but this could be a major error in your schedule plan. The amount of time editing and rewriting will take depends on things like who your client is and how your first draft actually goes.

Extensive rewrites could mean that the same amount

of time it took to create the first draft is taken during the editing and rewriting stage. In fact, the editing time could even take more time than the first draft creation.

Therefore, be wary of advice to only schedule five percent of your time for editing and rewriting and use your mind in a more practical sense, rather than just following what seems to be the status quo when you are developing your schedule.

The advice of professionals and the realities of work are definitely not always in sync. We all overestimate and underestimate things at times. This is especially true for those of us who've been doing something for many years who become comfortable with our basic ideas of how we do things.

This is exceedingly true of those who are more 'green' and are simply relying on the things they've been taught in class and who are portraying what college taught them as hard and fast rules. They'll learn eventually, but what is not certain is whether or not they will eventually be willing to outwardly acknowledge that the facts of life are somewhat different from that which their schooling taught them.

In some cases you'll need to include in the schedule some time for your subject matter experts to actually review the material you wrote that is directly related to the information they shared with you. Unfortunately, you will not always have this as an option, but when you do, it is good to take advantage of it.

Determine when you are creating your schedule whether this is an instance where you can do that and if so; be sure to allot time in the schedule for the SME review process.

Remember also that for any and all pieces of a schedule entirely reliant upon others to perform any kind of a task, or to provide any kind of a response, you have an unstable element in the schedule. You must recognize that there should be cushion around those time slots scheduled and that you have to have a clearly communicated deadline.

The purpose of the deadline is to encourage the expert to complete their task within the allotted time, but that is not the only purpose. It is for you as well because you need to be prepared to proceed without them if they do not complete their piece within a certain amount of time.

You still have a deadline to meet regardless of whether or not others involved do what you need them to do and you cannot wait forever. Be certain that you communicate this to your SME because even if it seems like an obvious fact to you; it might not be apparent to them.

Make it clear that you really need their review in order to ensure the best product, e.g. the most accurate document, is delivered in the end, but that due to your deadline constrictions, you will have to publish the document without their review if it doesn't come by such and such a time.

Sometimes people, no matter how intelligent they may be, simply do not connect the dots. You are probably not the highest priority in their mind for the various tasks they are working on in their life and work schedule, so just be mindful of this fact.

TIPS:

- **Do not fail to allow extra time between stages of your project.**

- **Do as much research as you must to thoroughly prepare for the writing process, but put yourself on a schedule so you do not get lost in the research phase and rush through the writing process too hastily. There are many elements involved in delivering a quality document.**

- **Even if your document is superb, you are likely to lose out on future work if you cannot meet deadlines.**

Planning | Meet the needs of your audience

There are dramatic differences in the way you will need to write your technical document based on who the audience of your documentation will be.

Things that are different depending on who your readers will be include:

- Types of words you use

    o Simple words verses more complex words requiring a higher education to comprehend

    o Use of industry-specific phrases and terms verses the lack thereof

    o Level of technicality allowable with wordings

- Length and depth of detail involved in each information set
  - Thorough verses less thorough explanations at which levels of granularity

  - Steps beginning from scratch verses steps starting at a level requiring comfort with initial process to reach that level of step comprehension

  - Heavy use of illustrations in explanations of content meaning, verses use of illustrations only used sparingly within an already drilled down section of the topic

- Does the document need to be easily scannable for fast-paced workers/readers who will only ever have the time to scan the document for specific points of interest, or can it have lots of lengthy paragraphs where one has to really sit down and read through each sentence in order to consume the information presented

- Are you attempting to obtain the attention of a friendly or unfriendly crowd, thus requiring the need to find a way to write the information in such a manner that will perhaps win over the audience so that they'll develop a desire to learn about the topic, or are you writing to an already interested audience that requires no goading

- How instructional verses informational the document needs to be due to the actual use cases for the document. If the audience is expecting to receive step-by-step instructions on how to accomplish a specific task, then only sharing a bunch of information about the task is not going to make the document usable for them and will cause them to have to search elsewhere for that type of information which they needed your document to provide them.

It also needs to be clearly outlined for these sorts regarding how they find exactly what kind of information they need inside of this document because if you have step-by-step instructions for them that are buried within a lot of other text and no clear indication on a quick glance that the actual step-by-step information is going to be available on such and such a page, then they may never realize that your document is actually one that would meet their needs.

- The formatting of the document on many levels is dependent on the needs of the audience, this is with regard to the text paragraphs verses bullet pointed notes and the actual size of the font and coloring of the font and even the use of graphics and types of graphics verses the lack thereof.

TIPS:

- **First and foremost: Remember your audience!**

- **Who the audience is does affect how you should format your document.**

How to obtain necessary audience information

Figure 6 - Artist: Nick Chagouris - Copyright 2014

Part of planning for your writing project includes learning more about your audience in order to meet the needs as described above. You will not always intuitively know all of these factors and will have to include this as part of your research and planning process.

When you are assigned this technical writing task, you should, in that initial assignment conversation, find out as many details regarding your audience as possible.

Additionally, when you are interviewing subject matter experts (SME's), include questions about the audience in those meetings if appropriate and possible at all.

Example: "Who do you feel would best benefit from this information?"

Example: "What are the unique needs, if any, of the people who you believe would be reading about this topic?"

Example: "In what ways do you believe former documents addressing this subject have failed to meet the needs of the audience?"

Typically, your subject matter experts would have previously had to do their own kinds of research on the topics you are covering. They will have had their own struggles in the past in acquiring all of the knowledge they were seeking.

The SME has not always been a SME since birth, so they have a poignant perspective to share regarding the ease and struggles involved in becoming an expert on this topic.

Learning about their process, including the challenges they faced in the past can provide you with very significant insight above that of others who might write on this topic and can enable you with the ability to create a much more useful document than previously existed.

It is very important that you think heavily about the users of the planned document. A well-written and informational piece that does not answer the actual questions or address the true concerns of the audience it was intended for is not worth the time or effort involved in its creation.

If you have the ability to contact any of your future readers this would be an outstanding edge for you in being able to create the best possible resource.

Ask the future readers things like:

- If you were reading a document on this topic, what would you expect to get out of it?

- What do you need to know about this topic?

- How often do you think you would refer back to this document if you had it in hand?

- Why would you read this document repeatedly or Why not?

- What would make a document on this topic the easiest for you to read?

- What would make a document on this topic the most enjoyable for you to read?

- How much time do you expect you will have to spend reading this document?

- What kind of task would you be involved in that would spur you to read this document?

- Can you give me an example of a technical document on a related topic that you found useful in the past?

- Why was it useful and what makes it memorable for you?

- Can you give me an example of a technical document on a related topic that you found was not useful in the past?

- Why was it not useful and what makes it memorable for you?

- If you got to choose exactly how a document like this would be formatted, can you think of any specific ways you would want it to be formatted?

- Do you find the use of pictures in a document to be helpful or distracting?

- How much step-by-step information would you expect a document of this sort to have?

- What would be the primary points you would be searching for if you looked through a document covering this topic?

- What physical location will you be reading this document in?

- Is the environment very conducive to reading?

- If so, what makes it so?

- If not, why not?

- How thorough would you expect a document like this to be?

- Do you expect to learn everything there is to know about this topic in this one document, or do you just expect to find a few useful pieces of information to lead you in the right direction?

TIP:

- **The subject matter experts are always more knowledgeable about the audience for their specific interest than you. Learn from them!**

Meeting the challenges of the diversified audience

Figure 7 - Artist: Nick Chagouris - Copyright 2014

One extremely important point to consider when planning out the creation of your technical document is the diversity of your audience.

Now, before we dive into the topic of meeting the challenges of a diversified audience, it's important to note that you might find yourself in a job such as an oil and gas technical writing job where you are managing the creation of specifications.

In some instances such as this, you are not going to be considering your audience, nor are you going to be worried about the diversities of your audience. You will have specific rules to follow and a specific template guideline to basically fill out. This is not necessarily the case with all oil and gas technical documentation creation, however, for certain specification writing it is.

In these cases your innovative ideas for meeting the needs of a diversified audience are not going to be helpful or welcomed and could land you out of a job. These specific types of technical writing jobs require one thing basically and that is that you follow the format.

You need to use the lingo and explain specifically what needs to be explained in the way these things have been explained for decades. People reviewing these specs are conditioned to see them a certain way and anything that goes outside of that formatting, structure, or language is not going to be useful to them, nor is it going to keep you in your job.

As such, this is one of the few cases where you need to consider your audience as just the audience that needs all formal requirements to be met as outlined by the Company.

Knowing simple things like the word "Company" is used with an uppercase "C" and so on are things you need to pay attention to. Keeping the document consistent in terms of definitions and word usage and providing the information exactly as expected on all levels is what is most important here.

If you can do this quickly and efficiently then you are a superstar technical writer in that group and need not attempt to find ways to diversify the information because any of that is already built in as much as it should be and your following the guidelines will match the need.

However, there are countless other technical writing projects where you will need to consider thoroughly the way in which you will face the challenges of developing documentation to aid the diversified needs of the readers.

Things to consider when developing an exceptional technical document that meets diversification qualifications include:

- Within what different settings will your readers be using the documentation?

    o Office reference
    o Classrooms
    o Basic users
    o Super users
    o Supervisor and management use
    o Group use verses individual use

- o Technical users verses end users
- o First language of users (e.g. American English, British English or any other language)
- o National or international use
- o Cultural differences
- o Gender bias

You may need to format your document so it can meet the needs of a variety of different types of users and within different types of environments.

There may need to be sections for informational and reference use, as well as step-by-step instructional use. If so, those sections need to be clearly defined, separated, and easy for the readers to locate as separate sections.

The language you use for the document is something you must identify up front before beginning your project. Whoever has assigned this task to you should detail specifically if you should use American English verses terms that British English users are more comfortable with.

Differences associated with location/region of your audience would concern more than just basic text words, but also time and measurement terms you use within the document.

It might be important for any notations regarding time to be in military time, but you won't know that unless you are told, or unless you ask. You can refer to previous documents to get an idea, but even if you see things done a specific way in the other documentation it is still highly important for you to verify the criteria for this specific document.

If you will have readers for whom English is a second language that may affect some of the language you use. However, be careful with any assumptions in this regard.

You may be surprised to discover that while English is

the second language for the readers, they are perhaps even more versed with the complex terminology due to their training than some native English speakers.

Be aware of the way in which you will have to meet the challenges involved in the specific type of diversification of your audience. This also relates heavily to the industry for which you are writing.

Be very careful never to assume you need to 'dumb anything down' for any audience because of cultural differences. You will end up looking like a very foolish technical writer indeed!

The differences involved in the level of detail and the types of terminology should be based more on the levels of education, exposure, and expertise of the audience, rather than on any cultural or foreign language differences.

In some cases you may need to include information that would be more pertinent to your national audience in one section, or as noted, and then also information more pertinent to your international audience in another section or as noted within each section.

Another thing you should be careful of is the use of "he" and "she" or "his" and "her" because this is something you can easily overcome and which can be a very offensive mistake should you direct the wording to one gender or another.

You might use the old he/she or him or her type of lingo, but it would be better generally to avoid that altogether and use words like 'they' or 'them' or 'the user' or other related words that work and prevent your document from displaying a bigoted and exclusive tone.

Additionally, ask questions of appropriate people in order to determine whether there may be specific additions you should make to the technical document in question in order to ensure it meets the needs of the international audience.

You will also want to keep in mind the cultural differences you may face during the interview processes. Using a tool called 'precalling' is one way to ensure you are meeting this challenge head on.

When you begin an interview, be certain to detail the purpose of the interview to the person you are speaking with. Encourage them to let you know if anything you say or ask makes them at all uncomfortable and assure them that your top priority is to make this interview a positive experience for all involved.

This will help you open the lines of communication in the most efficient way and least offensive way possible. It lets the interviewee know that you care about making them feel comfortable and makes it easier for them to tell you when you've crossed any negative lines.

Building a positive relationship with those you

interview will assist you over the course of your career and make the subject matter experts more likely to return your call or respond to your emails if you should ever require their assistance again in the future.

Business is in part about building relationships. Where you fall short in some areas you want to shine through as strong in others so that you can continue to grow as a professional and form strong bonds that will help all involved in being more successful across the span of time.

You cannot please everyone all of the time, but you can work to forge whatever positive relationships you possibly can so that you can keep moving forward and taking advantage of opportunities as they present themselves in the best manner possible.

Depending on the nature of your technical document, you could choose to specifically outline which parts are geared toward classroom use, or which pieces are specifically targeting managerial usage. You might have defined sections for the end user, or you might advise within certain sections that a more advanced user can skip ahead to page such and such or to a different chapter.

There are many ways you can successfully manage the development of a document that must meet the needs of more than one specific audience type, but you must identify those needs during the preparation process in order to ensure you do not waste a lot of time or mistakenly turn out a document that will not meet the majority of the needs of the entire intended audience.

This is all a part of the organization process as well and we will speak to that topic now.

TIPS:

- **Be consistent with terminology appropriate for the industry.**

- Take advantage of the opportunity to build your connections and business relationships.

- Always consider your audience, before you begin writing.

- Avoid gender bias in your work

- If you are composing your document in English, be certain to know whether you are writing for an American English audience or an audience in the U.K. for example as there will be minor spelling and formatting differences to which you should pay heed.

- Beware of explaining things at too elementary of a level if you are writing for a more experienced audience.

- If your readers will have little to no experience on the topic be very careful not to make your writing inaccessible to them by leaving out important details someone new would require to understand the majority of the material.

Organization | Save time and get the best results

Understanding the principles and methods of information management will be a great advantage to any technical writer. You will be gathering a potentially large amount of information over the course of time and will be creating many different documents. Some of these documents are writing drafts, while others are notes taken from your research and interviews, yet still others are documents you've downloaded or gathered together for the purpose of research in connection with each project.

When you are putting your document together, you need to be able to easily locate all of the information gathered for each specific topic; otherwise you will easily miss including points in your final draft that you worked hard to accumulate during your research and preparation stages.

You may choose to create a hard-copy organization system, or an electronic one. It is more common to do both, but likely best to simply create a digital version of any paperwork so that you can find everything in one location.

Make sure you have a back-up copy of all important electronic documentation. Do not take any chance of losing important interview notes, or hours of work because you spilled coffee on your laptop.

There are many cloud hosting solutions for media storage and you may prefer to keep all of your documentation stored in the cloud in order to ensure it is accessible to you wherever you may be; however, even with cloud hosted solutions, there is always a data loss risk, so come up with a secure system that includes a contingency plan, should any electronic failures of any kind occur.

Technicians and others documenting their own tasks

Technicians and other professionals must document their own tasks, results, experiences, and so forth for

113

the purpose of record-keeping, but also for specific sets of readers.

In these cases concurrent task organization is of particular importance. Rather than separating out your technical task work from your technical writing efforts, you would do well to include these together in the same step. If you wait until you are done with all of your work to begin the documentation and writing portion as part of a separate process it would be more than easy to lose track of important pieces of information.

Technicians must write in order to communicate the results of their technical work. The documentation technicians create is also often important supporting documentation for legal purposes.

One aspect of the organization process for technicians is the classification of documentation. There are different forms of technical writing and these may be categorized by function. Documentation of varying sorts may be classified by time period within their parent categorization.

When performing any technical writing as part of an employee role an especially important type of classification is internal vs. external as the rules associated with each vary dramatically. Security is of definite concern, but also reputation, branding, and confidentiality levels.

Determining the classification of the document is particularly important part of organization as each document type may have an entirely different format approved for use within each organization, as well as separate component inclusions.

Organizing these aspects of technical documentation will aid the technician in better determining how to perform parallel organization methods for the end of more effective, efficient deliverables.

File Naming Conventions

One major point of being organized is to ensure you can easily find what you seek without wasting much time looking.

Within some electronic systems, you can perform a search on your files by typing a word or two into a search field and results yielded will include any words from the name of the file itself, as well as any words from the text within the files.

However, not all systems work like this and the search feature may only work to show you files with names that have corresponding words, yet not any documents that have these words in the body of text within the file.

To be safe in terms of searching ease, as well as in simply scanning the screen when you have a folder opened revealing the contents of that particular folder, you need to use appropriate file naming conventions.

This means the name you choose when you are performing a 'save as' on your document.

Although you are able to see the last saved or last modified date when using the detailed view of your folder and file structure on your computer, you cannot always perfectly rely on this feature to tell you the date you want to know that is associated with this document.

If it is an interview file, for instance, you would do well to be able to remember the actual day, month, and year the interview was actually held.

Therefore, including a date in the file name of your document will prove useful to you as a habitual part of your file naming process.

If you are naming an interview file, it would be particularly wise to name it something like: 05022012-Marshall-Smith-Interview.docx.

Potentially you might now, or later, perform a wide variety of technical writing tasks for different companies and in different industries. Therefore, it could be even better if you named your file: 05022012-Marshall-Smith-CompanyZ.docx.

The company portion of the file name could be the name of the company you are writing for, or it might be the name of the company Marshall Smith works for. Either one will give you a good historical reference and scanning capability for expediently locating the documentation you require, whether by simply looking through the list of files, or by using a search engine.

Beyond the files themselves, it is wise to create a folder structure that is well named. Being careless with the naming of your files and folder will result in many future headaches.

If you have more files on your computer or within your document storage system than are pertaining to technical writing assignments, then it makes sense to create one overall folder on your computer for all technical writing tasks.

You can call that whatever you like, but you could simply name that folder: Technical Writing. However, if you do more for a specific company you are working for than just technical writing (This is from an independent writing perspective where you are working from your home computer.) then you might choose to name the top-level folder by the company name to which you are providing services. Within that folder you can create a technical writing folder and separately folders for the other types of jobs you do for that company.

Within your technical writing folder you will then create project folders so you can easily locate any and all files associated with a specific project within one project folder location.

If you are not utilizing electronic means for your document storage, you will still need to carefully organize your paperwork in clearly labeled folders, drawers, or other storage containers in order to easily locate the documentation you need both throughout the life of your project and in the future for reference and answering potential questions that may arise.

TIPS:

- **Don't let a time crunch ever prevent you from managing your documentation and notes in an organized, well-labeled fashion. You will pay for this mistake in the future if you think it is okay to be disorganized in order to save time.**

- **Use clear, easy to understand names for files and folders.**

Organization and protection in the real world

Figure 8 – Artist: Nick Chagouris - Copyright 2014

If you are working with a team of writers or other people who perform various tasks for projects, it is possible the company you are working for, or your project coordinator may use a cloud hosted collaboration environment where you upload your notes, versions, and varied documentation within that environment so all with appropriate permissions have access to see both the documentation and the proof of progress in order to monitor when various benchmarks are achieved, as well as to ensure the effectiveness of the team collaboration.

If one task is not on schedule then, other members of the team have the opportunity to seize that task and complete it so that collaboratively the project deadline will be met.

These systems are exceedingly useful and are becoming much more popular as companies are enjoying the removal of financial overhead via the use of remote workers and cloud hosted project management systems as collaborative team environments.

Some examples of collaborative team systems and project management systems currently available on the Internet are:

- Teamwork (https://www.teamwork.com)
- Podio (https://podio.com)
- Basecamp (https://basecamp.com)
- Wrike (https://www.wrike.com)
- Zoho (https://www.zoho.com/projects)
- AceProject (http://www.aceproject.com)
- Huddle (http://www.huddle.com)
- FreedCamp (https://freedcamp.com)
- ProjectBubble (http://projectbubble.com)

Be very careful here because these systems are software applications with faults of their own. Many newly developed systems are available for use today and sometimes have bugs in both the programming, as well as the concept of what is the most effective structure for managing remote team environments.

Older systems are also equally as capable of flipping your world upside down on a project because even though there may have been a great deal more time to vet out the actual code, this does not mean that an error will never occur. There are updates that can alter things and data storage issues that can arise to name only a few potential problems that can result in the loss of important data.

You may trust that a commercially available system has a backup plan to handle any issues; however, your trust can easily be betrayed due to a number of typical issues that occur within digital environments.

The point is not to avoid such systems because they can prove to be incredibly beneficial and may support the ability to more effectively manage remote teamwork than many systems of work established within brick and mortar offices and those used by well-known corporations.

The point is, rather, to make sure you have complete back-up documentation associated with every piece of work you do. Beyond this even, you may want to take screen shots of your collaborative environment during different points in each project in case you need proof that you did communicate and/or upload the expected deliverable within the required timeframe in case there is a glitch in the system and you find any action or work of yours comes into question.

This is the old CYA adage. This means cover your behind. People may have no ill-will whatsoever, but still question when appearances lead the belief that a team member has failed to accomplish her or his agreed upon tasks.

Particularly if you are working independently it is critical for you to protect your reputation as an honest, reliable worker and the only way to do this is to protect your work and save proof of all correspondences.

Retaining old emails for all work matters is also imperative because memories are faulty and people become overwhelmed and their personal systems of organization are often not fool-proof. Additionally, your team member or team leader may be tired or may not be sober when they are looking for information related to something you were accountable for and might snap at you thinking that you failed them in some way, when the reality is they simply forgot something like a request or a change or that a task was completed and you are not even involved in that stage of the project any longer.

Rather than having to involve yourself in any kind of emotional discussion or heated debate, you need do nothing more than forward the old email or proof of what actually occurred to the inquirer so they can then have the information they need and can decide for themselves at that point whether to apologize or just blow off their indiscretion and continue moving forward in the knowledge that you are a reliable worker who is an excellent resource for inclusion in future projects.

Another issue that can occur in these online collaboration environments is the lack of versioning. Often the system allows for changes by various users of the application to the uploaded documentation. If there is no tracking of who did what, and/or when changes were made, or whether there was a previous version, or a saved copy of that previous version, then your original document could be lost forever if you do not have both a back-up system and a well-organized system of storing your documentation apart from the team application.

For each feature of a program, development, coding, maintenance, and management are all involved. Some companies and/or individuals who create these applications will not include all of the features due to cost and time constraints. Additionally, they may simply be interested in meeting a more limited need, or requirement with this particular piece of software they are providing.

Do not assume anything in terms of storage, back-up,

and feature functionalities for any team collaboration environment you either choose or are required to use as part of a job you've accepted.

Now that we've discussed the writing preparation process, let's consider the research aspect more in-depth.

TIPS:

- **Always create a back-up copy of your work.**

- **Protect your reputation by keeping a cool head and being objective whenever tensions rise in association with anyone you are dealing with professionally. This will eliminate the potential for the 'closed doors' you would otherwise find if you've burned bridges or are reputed to be a hothead or difficult to work with.**

- **Familiarize yourself with the different types of collaboration and digital project environments**

available if you are planning to work remotely, independently, or on any contract assignments.

Technical writing research – methods of research

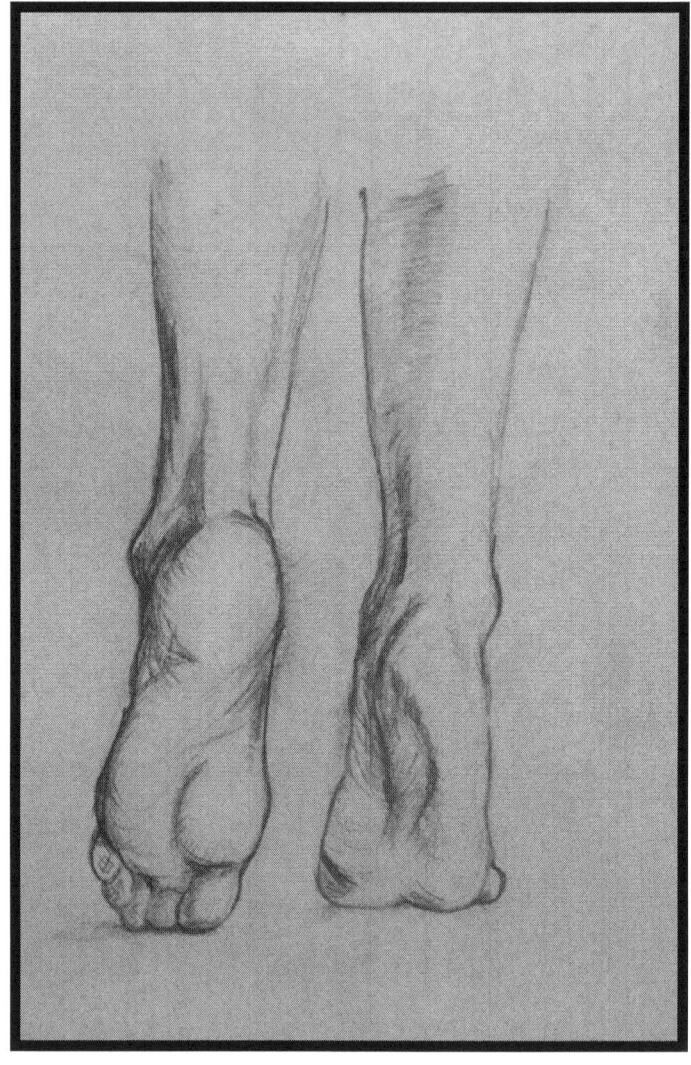

Figure 9 – Artist: Nick Chagouris - Copyright 2014

Different types of technical writing require separate methods of research. In some cases the technical writer is the SME (subject matter expert) where in other scenarios the technical writer works in a wider variety of genres and cannot possibly be the expert in all of the areas they write about. Even times when you the writer are the expert in the subject there may be forms of research you will need to manage in order to provide the full resource you are creating.

Research may include:

- Interviews
- Online
- Library
- Reference materials
- Working through a process
- Collaborative development
- Field work

Not every technical writing project is the same. Not

every end goal is identical with each technical manual, document, or book.

Always evaluate what you are trying to accomplish at the outset, as part of the formerly discussed preparation process, in order to determine what method of research you will need to use in order to develop the most appropriate deliverable for your project.

Two of the most commonly used words people use when discussing research are qualitative and quantitative. Personally I think people use these words a lot just to make others aware they know the words, e.g. to make themselves sound 'smart'. Still, these are good to know and important to understand if research is a part of your profession.

Qualitative research and quantitative research are two different types of research that support one another. Qualitative research can provide guidance in determining what type of quantitative research to perform. It may be focused on the 'why' consideration in your research. Quantitative research can provide support, or proof, to the assumptions or conclusions asserted from the qualitative research process. Quantitative research is more data-driven. One example of this is analytics where data is gathered to support a conclusion or theory.

Quantitative research is considered more rigid than qualitative research where like is specifically compared to like in terms of comparing, for example, a specific answer to a specific question or query of some kind with the exact same question's answered compared.

To increase the acceptability of qualitative and quantitative data, methodologies have been developed as part of the qualitative and quantitative research process.

The following links lead to some sources of information on structuring your qualitative and/or quantitative research process.

http://www.ccs.neu.edu/course/is4800sp12/resources/qualmethods.pdf

http://www.idosi.org/wasj/wasj12(11)/23.pdf

http://www.measuringu.com/blog/qualitative-steps.php

Conducting Interviews

If preparing to create a technical document associated with the development of a new technology there are often numerous people who have been involved in the development process. Taking advantage of any opportunity you may have to interview the individuals who participated in that project can be an exhilarating part of your work as a technical writer.

These interviews can provide you with the benefit of learning a great deal more about the new technology itself, as well as help you to understand what was involved in the progression of the project from the point of idea generation through development and completion. It can be inspiring to discover the many challenges those involved overcame in order to turn an idea into reality.

You can garner aspects of perspective that may not have come to your mind previously or ever even when considering the subject and the technology itself.

Details of the technology development process may enlighten you and ultimately your readers. Your work can become a much more thorough and even interesting document or manuscript by including information provided by various members of the development and/or business team involved in bringing the original concept to fruition in some way that may have mirrored the initial idea or that may have evolved and mutated throughout the development lifecycle.

In this type of scenario where you have the opportunity to conduct interviews associated with the development of a new technology you may want to ask the following types of questions of those you interview:

- What was your role in the development process?

- What was your vision when you first got involved in this project?

- How did you become involved in this project?

- What sort of experience had you with this type of technology prior to your involvement in this project?

- How did you come to work with the team you ended up becoming a part of?

- What was your favorite aspect of this process?

- What struggles did you face during the course of your involvement in this project?

- How did you overcome the challenges associated with this project?

- Where did the idea for this project stem from?

- Are you satisfied with the end result?

- Was there anything you would do differently if you had it to do over again?

- What lessons would you take away from this project and apply to the next one?

- Do you consider the technology to be complete or do you expect further improvements to be made as time progresses?

- What is a key point of interest you believe someone who should attempting to become an expert on this technology should be aware of or focused on?

- What are the different applications of this technology?

- What other technologies would integrate well with this in order to provide the most robust solution to meet varied challenges?

- In what ways do you believe the development of this technology will better societal or professional capabilities in the real world?

- Can you tell me about the person who initially involved you in this project?

- Will you dive into describing the detailed technical specifics of your area of expertise in relation to this technology?

Dependent upon the specifics of the technology itself, as well as the information you are provided in the interview, you will want to develop your own unique questions and ensure you do not entirely follow a rote set of questions you find in this book or in some online search. Instead, be sure to ask questions you had not previously written down that arise as you listen carefully to the person you are interviewing.

Your interview will be much more successful if your interviewee believes you are interested in what they are saying and actually capturing the information they want to share.

You should always record any interview you do if at all possible. This will help you to avoid missing important details and provide you with a reference to return to when you are reviewing your documentation. Additionally, you may need to provide proof that you actually completed the interview and the best proof is a recording of that interview.

If you are conducting a phone interview there are many different types of recording devices you can use, but one of these types is to arrange a web-connected phone call where your participant calls in through a phone number associated with that specific project and the web system records the entire telephone conversation.

It is important for you to download your conversation from this application after the interview is completed to ensure you have your own copy in case anything should happen with that web service. Additionally, you can easily share your downloaded sound file with others who do not have accounts with this specific web service without requiring them to create an account.

Before the interview begins, make sure you tell your interviewee you are recording the interview. Let them know that if you ask them any questions they are uncomfortable answering to please inform you of this so you can avoid insulting them and ensure they understand you are completely comfortable with a truly open dialogue.

You do not want your interviewee to feel like they are strapped to a chair, sweating, while an interrogator is giving them the third degree. This is one of the ways to aid you in building a rapport with the person to whom you are speaking.

Throughout the interview you should thank them at various points for what they've shared and definitely show your gratitude at the very end. You want people to be willing and happy to interview with you, not avoiding your phone call.

Make remarks about points they mention that are interesting and let them know this as part of the conversational ball passing, but beware to not sound fake in doing so as nothing hits the recipient of a compliment quite so much like a pit sinking in their stomach as does hearing someone clearly sending false accolades their way as part of a routine, rather than being a genuine recipient of the information they are sharing.

Carefully develop your reputation as a respectable interviewer in order to obtain a much greater amount of information, a higher level of quality information, and to improve the potential willingness of people to set aside time to answer your questions and keep your career afloat.

For people who are inexperienced at interviewing others it can be a bit nerve-wrecking at times. Still, for others they have to work to overcome feeling shy no matter the number of interviews they've managed.

Try to relax and become comfortable before and during the interview in order to reduce the amount of agitation experienced throughout the course of the interview. You may be benefitted by meditating prior to the interview. Take deep breaths before and potentially during the interview, but not at the expense of sounding like a creepy breather on the phone or in person.

Remembering and ensuring the importance of your participant's comfort to the best of your ability is critical.

If you are working in a collaboration application, share the results of your interview with the appropriate team members so others working on this technical writing project also have access to the knowledge you've acquired in the midst of your research.

Beyond writing about the development of a new technology, you may easily find yourself writing about any number of other technical matters as this is merely one example.

In some cases you will be writing about a long-established technology and will want to interview experienced users of the technology and potentially managers of teams who use the technology.

The questions you pose to your interviewee for the different types of scenarios should vary in some ways dramatically depending on what you are writing about.

If you are writing a new version of a technical document that was previously written and you have the chance to interview the writer of the earlier version, this too could greatly benefit you. Ask them about their research and how they originally developed the document you are working on.

Query your subject as to what ways they think their

subject matter may have changed, or evolved over the course of time to determine what areas they would personally choose to update at this point if they were completing the release of the new version.

Obviously this is not always or perhaps even often going to be an option for you, or even appropriate, but if you are in a situation where this is a viable option, take advantage of it to further develop your industry contacts, and to ensure you are fully aware of the details your document needs to include. Never arrogantly assume you have considered everything there is to consider when you are starting a new technical writing project.

TIPS:

- **Do your best to establish a rapport with your interviewee/respondent – This will make it much easier to gather more information than you would from an uncomfortable person who is just** hoping the interview ends soon.

- If the individual you are interviewing appears confused or gives other signs of not understanding the question you are asking, be ready to rephrase your question without showing any signs of impatience in the process.

- Be careful to avoid asking 'leading' questions because this will potentially limit the amount of information you acquire and in certain types of technical writing assignments can even be detrimental to your desired outcome.

- If you have an opportunity to hold the interview face-to-face, take advantage of that as it will increase your opportunity to establish a rapport, will increase the number of business connections you have who better remember who you are, and provides you with a more relaxed setting.

You also have the advantage of watching their body language in person and better identifying whether or not the interview is progressing smoothly or if you need to make a few adjustments in what you are asking.

Someone who is speaking to you in person is also often much more comfortable spending more time with you than they would be interested in spending involved in a telephone conversation.

- If an in-person interview is not possible, always choose a telephone interview over merely submitting a written list of questions to the person you need information from.

- Begin your interview with shorter, simpler questions if possible and progressively work toward the more complex questions, rather than startling your respondent at the outset with a heavily involved question.

- Never ask two questions at the same time. Give your interviewee a chance to respond to each question individually.

- Reiterate some of the statements your respondent makes in order to assure them you are listening and understanding what they are saying, as well as to ensure for yourself you are not 'getting it wrong'.

- Carefully prepare your set of questions prior to the interview. Do not 'wing it' and hope for the best. Your interview questions should be designed and developed as far in advance as possible, allowing a cushion of time for you to make adjustments and review. Consider your desired outcome (the kind of information you need) when developing your set of questions.

- Always review the questions you will be asking shortly before the interview so they are freshly in mind when you are on the spot. Not knowing what it is you want to ask makes you look unprofessional and unqualified for the interviewing task.

- Do not ask questions unrelated to the topic you are covering unless they are specifically designed to develop a better understanding of the person you are interviewing. You will distract and exhaust your respondent if you ask too many unrelated questions.

- Always be courteous and never show any signs of impatience or condescension during an interview. You should never make the individual you are interviewing feel like they are under any kind of pressure or are being judged for any of their responses.

- Make it as easy as possible to answer every question you are asking. You want to gather as much information as possible to ensure you ultimately deliver the best document.

- Ask open-ended questions so you are not leading your interviewee to simply answer 'yes' or 'no' as this will greatly limit the amount of information you are able to obtain.

Online Research

There is a great deal of good and bad information available online. Debates are held regularly about wiki sites and the value of crowd management in terms of whether or not data is refined to the point of credible accuracy via the volunteer efforts of the general public in association with these wiki's.

Take care when determining what information to use as part of your online research. Beyond the area of the wiki, there are also a large percentage of websites where every bit of information presented is merely the writing of a random individual who chose to type something and post it online.

With the ease of use and free access to CMS systems, where people can simply push a button and basically install a website online, nearly anyone can make their opinions appear to be official fact. Fake news stories are posted regularly and are a good example of the reason to take everything you read online with a 'grain of salt' or to double-check with a variety of sources when you are presenting information you've discovered online.

Even some very intellectual, respectable people I've known, physicists, lawyers, IT professionals, and others, have been duped by online sites that appear credible.

One way people perform research online as part of their technical writing work is to review what other articles and technical papers are available on said topic. Additionally, there are many books to view online through the use of the google books repository and via purchase in hundreds of online book stores.

Digital books are very popular now and have created an environment where buying books to read as part of research is much easier and faster. With a simple click to pay and download, the book you want is available without drive time involved.

From the comfort of your easy chair you can locate, purchase, download, and read any of millions of electronic books.

It would likely take too much time to read every book on the subject you are researching, unless there are not many available. However, scanning through a book for important pieces of information and perspective on the way others have covered the topic can be a very beneficial introductory step in your project once your topic is clear.

You must be very careful not to plagiarize anyone else's work, but learning from their work is an entirely different matter and can be very beneficial.

There are very helpful resource repositories available online as well in the form of statistical information you can gather, SPE (society of petroleum engineers: http://www.spe.org/publications/) papers, numerous white papers, data sheets, and case studies to assist you in pulling together technical pieces of data to include in your technical writing document or manuscript.

The following are some examples of a variety of different types of resource repositories:

AMSER: Applied Math and Science Education Repository (https://amser.org/index.php?P=Home)

OnCoRe Blueprint (http://www.oncoreblueprint.org/resources/repository_information.html) has a long, excellent list of repositories you can choose from in areas of math, science, computing, and additional subject areas.

TechRepublic has a white paper repository here: http://www.techrepublic.com/resource-library/whitepapers/

United States Environmental Protection Agency (EPA): http://www.epa.gov/airquality/oilandgas/whitepapers.html

Sparx Systems (Software Development) White Paper Repository:

http://www.sparxsystems.com/resources/whitepapers/

Embarcadero (Software related) White Paper Repository:

http://www.embarcadero.com/resources/white-papers

Dell Computers White Paper Repository:

http://en.community.dell.com/techcenter/systems-management/w/wiki/5103.dell-repository-manager-white-papers

NASA (National Aeronautics and Space Administration) Goddard Library Repository:

http://gsfcir.gsfc.nasa.gov/search/casestudies/View%20All

Emerald Group Publishing Case Study Collection:

http://www.emeraldgrouppublishing.com/teaching/casestudy_collection.htm

CIRTL Network of STEM Publications Listing:

http://www.cirtl.net/bibliography/keyword/CIRTL%20Publication&sort=author&order=asc

USA.gov Source for Statistical Data:
http://www.usa.gov/Topics/Reference-Shelf/Data.shtml

AERA (American Educational Research Association) Scientific Presentation Repository:
http://www.aera.net/Publications/OnlinePaperRepository/tabid/10250/Default.aspx

CoRR Computing Research Repository:
http://arxiv.org/corr/home

Forrester Research Reports:
https://www.forrester.com/search?N=10001+40004&sort=3

Gartner Research Reports:
http://www.gartner.com/technology/research.jsp

IDC International Data Corporation:
http://www.idc.com/search/simple/perform_.do?page=1&hitsPerPage=25&sortBy=DATE&lang=English&srchIn=ALLRESEARCH&src=&athrT=10&lp=1&lpr=3M&cg=5_1282&cg=5_1805&cg=5_1804&cg=5_1290&cg=5_1286&cg=5_1287&cg=5_1284&cg=5_1285&

cg=5_1291&cg=5_1294&cg=5_1293&cg=5_1288&cg=5_1289&cmpT=10&pgT=10&_xpn=false

Ovum Research:

http://www.ovum.com/research/

Yankee Group (IT Market Research):

http://www.yankeegroup.com/home.do

Aberdeen Research:

http://www.aberdeen.com/_aberdeen/-/-/-/search.aspx

Quandl is another source of many statistical data sources:

https://www.quandl.com/resources/data-sources

Be aware that not all of these resources are free and some are even incredibly expensive. If you are going to absolutely have to purchase a costly case study or research report, make sure you have this cost built into your overall project charge so you don't end up working for free.

It would even be wise to charge an up-front fee before

beginning your work if you are going to incur expenses during the research process.

No matter how big the corporation you are working for and no matter how reputable the firm you are assisting, you should never neglect to CYA and make sure your expenses are covered at the outset. It is also possible the company or individual you are going to be delivering the finished product to already has a membership with one of the online research resources and can simply provide you with a login credential in order to make the information easily accessible to you.

Ask these questions before accepting accept an assignment. This will prevent risking as much disappointment after you have already been engaged and are under obligation to whoever you are reporting up to.

At times you will be getting online to access the personal repository of the subject matter expert's notes made available for you so that you can turn their doodles into a completed document.

Many colleges have excellent online libraries and document repositories you may be able to access. Here are a few examples:

Virginia Tech University Library (http://www.lib.vt.edu/find/databases/O/opendoar-directory-of-open-access-repositories.html)

University of Washington Bothell (http://www.bothell.washington.edu/learningtech/instructional-resources/repositories)

Cornell University Library (http://arxiv.org/)

University at Buffalo Libraries http://library.buffalo.edu/

Illinois Institute of Technology Ethics Education Library http://library.buffalo.edu/

Stanford Law School Case Studies https://www.law.stanford.edu/organizations/programs-

and-centers/environmental-and-natural-resources-law-policy-program-enrlp/case-studies/stanford-law-scho-3

Purdue University Library
https://www.lib.purdue.edu/

Harvard Library
http://library.harvard.edu/

Duke University Libraries
http://library.duke.edu/

University of Michigan Online Library
http://www.lib.umich.edu/

There are of course many, many colleges that offer online libraries and the above list clearly does not display them all.

Professional and nonprofit organizations have also created excellent resource repositories available to the public and available to members only. There are many reasons why becoming a member of some of the different organizations can be beneficial to you with research resource accessibility being only one of those.

One good example if appropriate to your industry is IEEE (ieee.org). While they are well known by those involved in the field of electronics, they are also well known by the computer industry as one of the grandfathers of organizations established to connect professionals and provide valuable resources. There are actually many branches within that organization, as well as many others where you can locate credible information and even potentially historical data associated with your topic.

One IEEE educational resources can be found at http://education.ieee-cis.org/ .

The main IEEE site is: http://www.ieee.org/index.html

The IEEE digital library is currently called Xplore and can be found at http://ieeexplore.ieee.org/Xplore/home.jsp and you can choose to search via a variety of methods at this point in time, including by author and publication, by terms and there are also advanced searching options.

Sometimes online research can help you in simply locating the correct terminology to use in your documentation, or even formatting rules associated with the industry you are writing for. As you may be writing across multiple industries, it is always good to remind/refresh yourself rather than allowing yourself to fall into the habit of writing in a style appropriate for a previous job you did, but not appropriate for your current task.

Online research is typically performed by attempting to locate information you need via the use of a search engine and by trying a variety of key phrases that will help you locate the listings required.

If having difficulty finding something you are seeking online, try to think of more key phrases that would be related to the topic. If your searches continue to prove fruitless, you may need to get some searching help.

Using a key phrase tool can provide you with ideas regarding what words to use in your search by showing you a variety of related words associated with the words you are thinking about. By typing the suggested, related words into a search engine your luck in locating information on your topic may increase.

This can also help you if you are writing any technical document that will be posted on a web page as including related key phrases in your text can help search engine algorithms better identify your text as an answer to a search inquiry when you make it clear via the use of words already recognized as associated with the specific topic about which you are writing.

If you want the search engine to show results that only match the exact words you are typing, the use of quotes ("word") around your word will help narrow the search results list to a certain extent.

Google's advanced search tool can also help you with this: http://www.google.im/advanced_search

Google

Advanced Search

Find pages with:		To do this in the search box.
all these words:		Type the important words: tri-colour rat terrier
this exact word or phrase:		Put exact words in quotes: "rat terrier"
any of these words:		Type OR between all the words you want: miniature OR standard
none of these words:		Put a minus sign just before words that you don't want: -rodent, "Jack Russell"
numbers ranging from:	to	Put two full stops between the numbers and add a unit of measurement: 10..35 kg, £300..£500, 2010..2011

Improving your searching skill is an important aspect of speedily finding what you need via online research. Think of words that would actually appear in the text of the source you seek.

TIPS:

- **Be patient with yourself and the applications you are using while performing online research. This will help you yield the best results.**

- **Exercise your creative mind when thinking of what words to use in your searches.**

- **Save money if possible by spending a little more time searching before you purchase an expensive research report.**

- **Remember that thorough online research takes time and be certain to include that time cushion into your initial project plan/schedule.**

Library Research

In line with the previous section, first it is important to note that many libraries also have an online library catalog. (*Note as per the previous discussion of using American English vs. British English: The word 'catalog' can be spelled as catalog or catalogue in American English, but is spelled only as catalogue in British English.*)

The online library catalog generally serves the purpose of discovering whether a specific library has a specific book or journal.

You can search on the subject initially as long as you know how the subjects are defined in the online library catalog. You may need to get assistance from a librarian or look at a subject manual, but often the library websites are now set up so that you can click a drop-down box to display a list of the subjects and drill down from there.

The format type (audio or print) and which associated physical library location are options you may be able to choose when using the online system to request the book be held for you to go pick up.

Books listed in the online library catalog have been entered into a database. Therefore when you are searching for a book you are generally limited in your searches to the specified fields the cataloguer has populated. This means you are unlikely to be able to locate the book you seek merely by typing any word or phrase into a search field pulling data associated with various texts inside the book you seek.

Rather, your search must be associated with things such as author, title, subject, and potentially there may be a connection between words the cataloguer entered in as part of the description field, but again, this is limited.

One benefit of using a library for your research may be related to the ability to find historical newspapers, periodicals, and journals. It can be very informative if you are able to detail timeline progress of a specific technology and certain aspects of the history behind the technology's development.

Extensive research is also possible via the use of bibliographies available at the library. In these you can dive much more deeply into a topic by searching far beyond just the one book.

Use the bibliography reference tool to review many directly associated works previously used as support and research tools for the book you have thus far reviewed or are aware is describing the appropriate subject.

Each document, journal, book, or other medium has to have a certain focus and can only include a somewhat limited amount of information on the topic.

Through use of the bibliography you may even be able to find data sources used to draw conclusions and view other areas of related data as part of your own technical manuscript development process.

In order to comfortably find what you need in a library it is good to understand the type of classification system used by libraries.

In America, libraries typically use either the Dewey Decimal or the Library of Congress systems, while in the U.K. the Universal Decimal Classification is one of the standard ways libraries classify and categories.

If you want to learn how to use the Dewey Decimal system to aid you in your library research, this link takes you to a helpful instructional tool in how to use the Dewey Decimal system.

http://mcpl.info/childrens/how-use-dewey-decimal-system

The following link provides instruction on how to use the Library of Congress system: http://library.csun.edu/ResearchAssistance/LCC

One of the best sources for learning about the Universal Decimal Classification (UDC) system can be found at the UDC Consortium:

http://www.udcc.org/

Occasionally you may want to use library resources other than books for your research. Videos and audio CDs may be excellent resources on particular topics.

TIPS:

- **Sometimes the fastest way to become comfortable performing research in a library is by simply asking a librarian for guidance.**

Reference materials

Another form of research is assumed in some ways and included in the online and library research sections mentioned above, but here we will just delve a little more deeply into the process of obtaining information you require from a variety of reference materials.

As this book discusses indexing, understanding how to use an index and really comprehending the usefulness of an index will make your research performed via the use of reference materials a much more efficient and satisfying process.

Indexes are particularly helpful in guiding you to each and sometimes every reference to a particular aspect of the subject you are most interested in within a particular reference guide.

In many reference guides you will find a list of sources identifying books that provide additional information associated with specific areas of the book you are using as a guide.

Your research can become extensively in-depth when you study not only the reference material you originally find or are guided toward, but also all of the other reference books and publications noted in the original document you've reviewed as part of your research.

Remember you never want to copy and paste from one reference manual into your own technical documentation unless you are specifically just taking a quote and making it clear within your own documentation precisely from whence that quote came.

When you are writing for a company, they own the work you create. You never know if they might publish something publicly and if so they are legally liable and so there should be no plagiarism.

TIPS:

- You may include details about the reference materials you used in the bibliography.

- Using reference materials is one opportunity to really 'dig' for information you need to support the validity of your documentation.

Working through a process

Performing research by working through a process is a very effective method for specific types of technical writing. For instance, if you are writing about a programming language in an instructional manner, you may want to include code related to the topic, or even teach someone how to code.

By doing the work yourself and documenting each aspect and step of the process itself you can create a more flawless technical document because you have created and recorded while creating; also verifying the accuracy of your work and claims in the midst of technical document development.

This can be true as well of other types of technical writing not related to computer technologies. For instance, rather than using a case study as your source of research, you may be writing a case study. While working through the process of developing the case study you may be documenting what you are discovering based on actions you are taking and observations you are making about people, materials, or properties involved in the case study.

TIPS:

- **Verify the accuracy of your work.**

- **You are less likely to miss a step when you work through a process and document your experience.**

Collaborative development

Research does not have to be performed in a vacuum. You may be a part of a team developing a technical document, or merely a team of researchers.

There are many ways in which collaboration can be quite useful as part of a development and research process.

It may be possible to collaborate with others in gathering various pieces of information required to complete the research stage. This collaboration may be accomplished through the use of an online collaboration tool or project management tool.

Simply talking via telephone, comparing notes verbally, and emailing notes for merging may also be a part of collaborative development in research.

At times it could be required for you to travel to a mutually agreed upon meeting spot in order to collaborate with part or all of your team in person through the merging of records and the volleying of ideas in order to achieve a collaborative evolution to achieve more throughout the research phase.

(*Please note the way I use the words 'stage' and 'phase' interchangeably. In the past projects were said to be broken down into phases. During the early 2000's the appropriate terminology to use was changed to 'stages' in the oil and gas industry. However, not all adopted this change and most couldn't care less one way or the other whether someone said the word stage or phase. Therefore, in the real world of industry, using the words stage and phase when describing the project management process are synonymously accepted norms. Yet, if you were writing a technical document, this is a prime example of the need for consistency.

I have used them both in this book because I want readers who are not familiar with projects to learn both terms. But in a technical document this is a perfect example of how you would need to pick one (phase or stage) and stick with it throughout the entire document. You cannot assume every reader understands these two words mean the same thing.

A reader may end up wasting time attempting to discover through the pages of your document what you meant when you used the word stage vs. what you meant when you used the word phase. Many readers may believe that if you meant the same thing you would not have changed the word as for them, this is unnecessarily confusing behavior.

Just use phase or just use stage in every part of your technical document when describing a project. Never use both within the same document. Develop your own set of standards as they agree with the preference of your client or employer.

One engineering manager will insistently demand you only use the word phase, while another will demand the opposite, and others won't care one way or the other. Find out from whomever you are reporting up to or from whoever is paying you, whether or not they have a preference and if so, just what is their preference for now and future reference.)

Field work

In some cases you will need to get your hands dirty as part of your research.

As depicted earlier in this book, technical writing may be defined as writing on a specialized topic. Part of the research involved in this specialized depiction or instruction may either include the need for travel, or may be associated with the very act of traveling itself.

Field research may include observational study. Observing without influencing the object of your attention is a form of research. From there you must record what you see. What you see may be just one piece of the overall puzzle.

After recording your observations, then ensure your documentation is secure and organized as discussed in the previous organization area of this book. Later you will analyze how the results of your observation connect with other pieces of data associated with your research and information gathering.

While observing your subject you'll record any potential behaviors, physical characteristics, and possible changes or alterations. Details of your notes are dependent on subject matter and research purpose.

Interviews may also be included in field work. Travel may be another aspect of this research.

Visiting an operations center, a chemical plant, a shipyard, a construction site, an offshore rig, a manufacturing plant, a variety of industrial settings, or other areas where work associated with your topic is being performed are included methods of field research.

In the energy industry there are locations where equipment is being set-up, built, managed, and operated; where processes are developed, flowing, and transformed. Each piece of equipment, from small tools to large machinery components may be inventoried through the use of a tagging system.

Some technical documentation requires notations of tag numbers and/or the collection of tag numbers. These tag numbers may also be electronically recorded for you to gather without going out into the field. However, it is also possible it might be a very good experience for you to actually view the equipment and operations materials involved with the work you are writing about.

Field work may also involve some form of participation that could be associated with the working through the process area of research mentioned earlier. The participation could include joining in with participants in a study, gaining a more intimate view of interactions, causes, and effects.

Observing and participating in discussions may be a part of the field research you will be performing and you may want to use a recording device with this area of field work, as well as in other areas of field work formerly described.

TIPS:

- **Field research is an active form of research where you go on location to get direct, visual information.**

- **Field research can be quite fun!**

Technical grammar and writing styles

Many mechanical grammar basics are going to be the same regardless of the type of writing you are involved in. Many professional writers are of the opinion that some grammar rules are meant to be broken. Many buck authority as to what proper grammar actually is, while others simply bow to different authorities.

One well known rule of difference in terms of punctuation and paragraph structure is the use of the space between the end of a sentence, directly after the period, and the beginning of the next sentence.

The rule is always: If you are writing for a print publication then there are two spaces between each sentence. If you are writing for an online publication you must only use one space between each sentence.

You may read any number of reasons behind this rule, however, one very practical reason this rule was established was because of the effect two spaces between each sentence had on the appearance of content on the screen for readers prior to more recent technology updates. Just a few years ago, if your text included two spaces between the period at the end of one sentence and the start of the next sentence, the rendering system of the browser, particularly in a mobile device, would insert additional characters, making the text look messy and difficult to discern in the areas where the double spaces were located.

That is still a problem that happens on some devices and within some browsers, but it is not as common of a problem as it was in former years as programmers have worked to remove this issue. However, the rule for online publication still remains that you should only put one space in between the end of one sentence and the beginning of another.

You may occasionally still find someone who is ignorant of this rule of online writing and to whom you have to turn work into for their review and approval to post online. The last time I had one of these negative kinds of discussions was in 2009 when I was doing some web content writing for a lawyer. He was completely convinced of his own level of knowledge despite having any technology training of any sort. He simply could not stomach the idea that there should be one space instead of two.

If you are unfortunate enough to come across someone like this you may handle the issue in a number of ways. To preserve your reputation and ability to continue to be sought after for work though you should be careful not to treat the ignorant person as if they are as naïve and hard-headed as they actually are.

You have the option of calmly explaining to them the differences between writing for print vs. writing for digital publications. If the writing you are doing is going to be transformed into a pdf copy though, you can simply give in to their demands if you choose. The best thing to do though is to let them know that you are not 'green' or without knowledge or experience, and thereby are worthy of continuing to hire for this sort of work.

The way you can do that without embarrassing them too badly is to calmly reference the differences between print writing and Internet writing and email your client a few links to different sites online that explain the rule of using one space instead of two. Of course there are many opinions and strongly held beliefs now that you should absolutely never use more than one space after a period, even with writing that will be published in print version, rather than online. Here are a few links you can use to show you have not just dreamed up this idea all on your own:

http://www.writersdigest.com/online-editor/how-many-spaces-after-a-period

http://www.chicagomanualofstyle.org/qanda/data/faq/topics/OneSpaceorTwo.html

http://theworldsgreatestbook.com/how-many-spaces-after-a-period/

http://www.slate.com/articles/technology/technology/2011/01/space_invaders.html

http://www.westminster.edu/staff/nak/courses/spaces.htm

https://owl.english.purdue.edu/owl/resource/560/24/ (This one claims two spaces, but is not speaking directly to the issue of writing for text online.)

If you do your research, you discover the Chicago Manual of Style and the APA style guidelines are contradictory on some points, including the space after a period/space between sentences issue.

However, we are talking about technical writing, which is not generally writing for the press, therefore would not follow APA guidelines. That being said, your technical manuscript needs to follow the guidelines of the company you are working for. Company policies and standards are the gods of guidelines in technical writing.

Sometimes you will be providing work for people who themselves have no technical writing background whatsoever and who recognize they do not know all of the rules involved in technical writing. In these cases you have the ability to 'set the rules' of the technical writing standards within your organization. As such you would be wise, as part of protecting your overall reputation as a skilled technical writing/technical communications professional, to adhere to the sort of grammar, punctuation, standards, and conventions as are commonly accepted in your specific industry.

NASA published a technical writing and technical editing grammar, punctuation, and capitalization handbook that can be found online here:

http://ntrs.nasa.gov/archive/nasa/casi.ntrs.nasa.gov/199 00017394.pdf

Studying *The Elements of Style by Strunk and White* can also be helpful. Many details in this style guidebook will help you deliver a clean, precise technical manuscript.

Books like Strunk and White's provide great tools for double-checking your grammar and writing style before you consider the document completed. A short list of unnecessary words to be avoided is included in *The Elements of Style* and the need to remove these words is applicable to any writing genre.

Words and phrases listed include:

- the question as to whether
- there is no doubt but that
- this is a subject which

Use the search function of your Word document when reviewing it, to identify and remove word sets well known as unnecessary blather.

TIPS:

- **Grammar and writing styles differ in field and industry.**

- **Associated Press style is not generally correct for technical writing unless you are writing a technical article for a newspaper or related type of media.**

Technical writing standards and conventions

Standards and conventions required for use by the technical writer may include such things as the industry terminology described earlier in this book, such as capitalizing the word Company. The formatting of the document and the parts to be included in the technical document may also be a part of following standards and conventions of technical writing.

Maintaining more consistency than a writer of fiction would in terms of specific word usage could be considered another example of a standard or convention of technical writing.

The importance of avoiding ambiguity is associated with this as well and is also considered to be an aspect of falling in line with the standards of technical writing.

Standards and conventions of technical writing have been carefully established in varied industries and as overall rules of technical communication in the effort to ensure the technical documentation is much more effective in achieving its goal than it would be otherwise.

Review technical documents associated with your assignment and in your industry to be certain whether or not you should include an index as part of the technical manuscript you are creating.

TIPS:

- **Conventions and standards are industry and even Company specific.**

- **Study your client or employer's standards prior to beginning your task.**

Formatting technical documentation

Technical documents may include parts and sections that other types of documents, journals, and books do not. The technical writer may need to include titles, a bibliography, different kinds of lists, an abstract, an appendix, an index, an introduction, a conclusion, and many other parts.

These various pieces of a technical document need to be placed in specific locations within the book or manual. For example, the index is usually supposed to be the last part of the document. This is one part of the formatting process involved in technical communication.

One thing not yet covered in this book is the process involved when you actually have to include pictures in your technical writing assignment. If this task does not fall on someone else's shoulders, it is important to bear in mind the copyright issues associated with images.

You cannot simply grab any photo from any website and use it in your technical document.

You must have permission to use any pictures you yourself did not create. It is possible to purchase images from a variety of stock photo locations online, as well as from hundreds, if not thousands of different photographers.

Generally a company will have their own set of images you can use to coincide with the technical document you are creating. If you are working for a company with policies and procedures, such as a in a larger corporate environment, you may need to pass the images on to the branding manager or legal team for approval.

Rules around images within documents that will only be distributed internally may be a bit more loose, however, you could still be liable if your document ended up in the wrong hands, so this is one very important area where you would be wise to think of the CYA adage and err on the side of caution.

It is never worth it to include an image you do not have the rights to use just because you want to do so or are pressured to do so and that point would be painfully driven home if the worst case scenario should occur and you end up receiving a big, fat fine to pay.

Most companies also have branding rules around images if those images are in any way associated with a logo of some sort. There are generally always rules about how much space must be left around a logo image. Ask the appropriate person or get a copy of the company's branding guide if you may be including a logo image.

TIP:

- **Be humble enough to realize there are likely organizational rules you do not know from experience or instinct. Ask questions!**

Indexes and references in technical writing

You may need to create an index that you would place in the back of your technical manuscript before your work is completed. Your need for an index depends on the nature of your technical writing assignment and the length of document you are creating.

References in technical communications may also be included in the form of a bibliography and/or footnotes within your document.

Where appropriate, create a thorough index noting the pertinent statements included within the body section of text in your document.

The index can make life much easier for your readers who are seeking out specific types of information by providing them with a tool to determine which areas of your book or other form of technical documentation are most specific to meeting their needs.

Within the entries of your index you may choose to either match aspects such as capitalization to that exactly as it appears in the body of text, or you may choose to capitalize the first word of each index entry.

When the initial entry is followed directly by the page number location/locator, use a ",' comma to separate the entry text from the locator.

Index entries should be short and concise as a small grouping of words to direct the reader to the actual subject or discussion within the body of text.

Recognition that your technical document or manuscript may in the future be used for someone else's research should aid you in comprehending the importance of creating an index as part of your work.

Potentially an employer or client may require you include an index in your writing, but it is also possible you may need to realize this need on your own as you are trusted to be the expert in how the technical document is developed.

In some cases, as in the case of short technical articles to note one example, you will not need to and should not create an index.

Alphabetizing the entries in your index is a standard way of formatting the index; however, you may be required by your employer or client to put your entries in chronological order. If you are directly assigned to create an index, check with your client and view historical company documents to determine which method to use.

TIP:

- **An automated index creator does not always do the best job, but in some cases it is sufficient and can save you a great deal of time. Test out your word processing program's auto-index feature to see if you can save a few minutes or hours.**

Editing technical documentation

Just as there are writing rules specifically for technical communication; there are also editing methods for technical writing. The editor must be fully aware of the rules of technical writing in order to accurately edit the document.

While you certainly want to keep the same rules in mind whether you are writing or editing, throughout the editing process you are heavily focused on the process of editing, whereas while writing the focus is on the 'meat' of the content development; that is, getting the data and information and instruction onto the page, into the appropriate sections in a congruous and professional manner.

To provide a thorough and comprehensive editing effort, it helps to separate the editing process into steps, focusing on distinct parts of the editing work separately.

This way you are able to minimize that which you are paying attention to and to achieve a more precisely edited document in the end.

Particularly when editing your own work apply the step-by-step technical editing methods to refine your perception and obtain more of an objective lens through which to view your text while looking at each individual specific piece individually.

While you might be (and are very likely to be) breaking the rules and performing edits while you are in the midst of the writing process itself; it is critical for you to schedule the editing process its own allotment of time as a task unto itself.

The official editing work does not begin until you have every section of your technical book or document completely populated. This means you are not officially editing until after you are officially finished with the writing stage of the project. This may sound foolish because of course you may be writing while you are editing, but you should not allow yourself to think of your text as having been edited until you have first completely finished writing the entire first draft from start to finish, and have then performed the step-by-step editing review as a siloed section of time both in real time and as you should have built into your original schedule and project plan.

It is entirely up to you whether you want to include steps such as finding and including images (if this is part of your job) before or after the editing process.

The purpose of editing a technical document is much more than just making it look pretty, polished, and professional. Remember that you are writing about a specialized topic and you are writing this for the purpose of informing or transmitting a concept through your text to your reader.

Therefore, the importance of precisely, cleanly, and clearly transferring this information is paramount to the success of the project of technical communication.

If your technical communication does not earnestly enable your reader to develop a deeper understanding of the specialized subject written about, or if it is does not enlighten the reader to help them attain an entirely new understand and/or grow a completely new knowledge set on topic, then the time, energy, and money spent is without value.

This is not to assume you can ever reach every single reader who ever picks up a copy of your document with absolutely perfect precision because the reader has to be of at least some sort of mindset capable of absorbing information.

However, you cannot use this excuse as a general rule. Your focus should be on expertly communicating the specified technical matter with the appropriate audience. (Remember the earlier 'know your reader' notes.)

There are many ways you can 'muddy the waters' and make the information you are presenting much more difficult to absorb and retain. One way is to be lax on the technical editing aspect of your job.

Mistakes such as typos can distract the reader and even worse can teach the opposite information your document should be delivering and can easily confuse your reader and prevent the technical document from accomplishing said purpose.

It is the editing process itself that really makes the document an acceptable and useful deliverable.

Earlier in this book the importance of organization when writing was stressed heavily. Now it is critical to note the vital aspect of organization during the editing process. Without an organized method of editing, the hope of transforming the technical document into a precisely instructional and/or informative work is weak at best.

If you are performing the technical editing work involved with a document another writer developed, rather than one you yourself crafted, your organized method of delivering your critique notes is imperative in aiding the writer to fully understand and apply the edits and notes you've included as part of your review.

As the technical editor you need to clearly communicate the concepts you are attempting to transmit to the actual writer in order to ensure they have the ability to deliver a technical document that will clearly communicate its proposed concepts to the readers.

When you first review a completed technical document to begin the editing process, review details such as whether all of the required parts of the document are actually present. You would be helped to have a page of steps involved in your editing process so you may check off each of these steps as you work through thoroughly editing the technical document.

If an abstract is required, does it exist?

Is there a table of contents?

If necessary, is there an index?

If a glossary is required, does it exist?

Are page numbers present on the document?

Are there headings in all of the appropriate places?

If tables are present, do all of the tables have headings and/or titles?

This first stage of the editing process as described above is merely reviewing the checklist of what parts should be in the document to ensure they are there. If any are missing, note that in an easy to identify/understand fashion.

You may next read through the text of the document. Check to discover if any inappropriate language is used. Are any statements within the content too abrasive, harsh, or judgmental?

Does any text or image in the document violate brand standards of the company for which the technical documentation is being created?

Next you may review the content to specifically look for any references that are not included. If one area of the text promised to discuss a specific topic later in the document, does that occur? If there are text citations for a table, appendix, figure, reference, or footnote, do those actually exist and do the ones that do exist all have text citations?

If a note details something will be included in the document, is it actually there? Do the page numbers noted in the table of contents match their related areas in the document? Do the locators in the index match the actual pages where the entries are discussed?

TIPS:

- **Remember editing is a different job and must have its own time scheduled in the project.**

- **Break the editing process into steps where the focus is unique to each step.**

- **Editing is not the same as proofreading, but these are combined in the project schedule.**

Reports

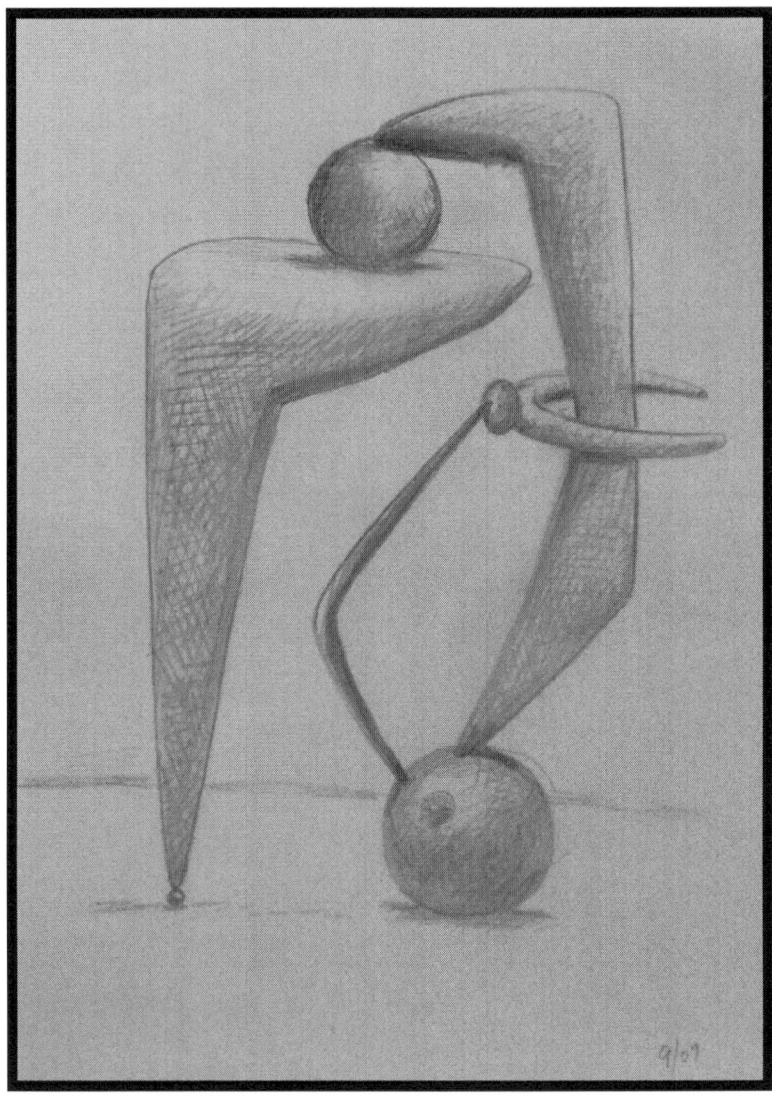

Figure 10 - Artist: Nick Chagouris - Copyright 2014

Reports are one of the more common types of technical documents created on the job. The University of Notre Dame has an educational archives website with a section in it called William Whitaker's Words (http://archives.nd.edu/words.html). That site provides the ability to translate words from Latin to English and English to Latin. This can be a good way to determine a meaning of a word for a more in-depth understanding of the purpose of a word.

The word report has a few different, but similar, meanings and in the Notre Dame dictionary there are about 20 different Latin words associated with the word report. From that dictionary the Latin words refero, referre, rettuli, and relates are closer to the type of meaning associated with reports in technical writing. The meanings there include report (on), bring back news and record/enter.

Other sources note the Latin word for report to be reportare, renuntio, and renuntiatio with associated meanings such as announce, report, return, declare, and give notice, or to return with observed information or investigated information.

A report is typically created in business to provide information to appropriate personnel and for the record so the proper supporting evidence and summaries of the supporting evidence provide decision makers with a substantial way to determine whether to take some form of action. Details in reports may aid others in more technical aspects of their jobs and provide an assortment of educational information to pull from and piece together the data they require for performing some job function.

The following is a list of many typical types of reports people are required to write.

Progress Report	Test Report
Research Report	Activity Report
Expense Report	Lab Report
Design Report	Complaint Report
Sales Report	Feasibility Report
Investigative Report	Accident Report
Visit Report	Trip Report

There are many other types of reports commonly required by office and field professionals in the workplace.

Examples of reports that should have a reader-focus (e.g. written for the reader vs. the record) are different types of reports that fall within the category of investigative reporting. Field reports and feasibility reports are two such examples. Investigative reports or other reports with a reader-focus should be written using a direct or deductive approach.

With the deductive approach, the main points are presented initially, in order to provide the answers sought by upper level management and others without time, need, or interest in scouring through the details. Only after presenting the main points should the writer then delve into the details supporting and clarifying the findings.

It is very easy for writers to neglect the reader's needs and fall into the bad habit of using the inductive method in their report writing. Using the inductive method for writing may be sufficient when writing only for the record and not for readers. However, it is much more common to expect there will be readers of your writing.

With the inductive or indirect approach in writing, details and explanations are presented first. Writers often want to explain themselves before they draw the conclusions, but this is more appropriate in creative writing than it is in technical writing that falls into the category of report writing.

Determining what is preferred by your employer or client though is your best method when you are working as a freelance or contract writer for an individual or company. There may well be copies of other documentation written in the style or using the methodology preferred by the one paying you to write that you can use to determine whether inductive or deductive writing suits the purpose of your assignment.

To clarify inductive vs deductive further:

Deductive: Direct approach – Not necessarily organized the same way in which the work was performed – Starts with the general and proceeds to the specific details.

Inductive: Indirect approach – Not necessarily organized the same way in which the work was performed – Starts with the specific details and proceeds to the general.

The general refers to the overall and main points.

Report writing should typically have the summary, conclusions, and recommendations in the beginning. You can liken this to the abstract in technical writing.

TIP:

- **Once you know the specific type of report you are supposed to create, there is often a template you can use to make the job easier. This may be an empty template or just a copy of a previous report.**

How much should you charge as an independent technical writer?

Figure 11 – Artist: Nick Chagouris - Copyright 2014

It's very important to not be so eager to get a job that you undercut both the competition and yourself. Remember that every aspect of your work is part in building your reputation as a technical writer for hire. If you charge too little you are setting an expectation and preventing yourself from earning enough to make the work worth your while.

While it's often tempting to charge a flat rate for any kind of remote work, it is not advisable. Clients often require changes, change their mind about project directions, and add tasks to the list of things they want you to do.

Choose between charging a per word rate or an hourly rate. Make sure you take into account the fact that more time goes into the work project than merely the first draft writing process. Build that into your cost. If there will be expenses involved, do not forget to include these in your charges. You can charge for expenses separately, or you can pad your per word rate or your hourly rate to account for that.

Earn trust by being as transparent as possible, but do not get stuck in a situation where your client is questioning your every expense and limiting you from getting the best job done as a result.

Do not begin the work until a contract agreement has been signed off on by your client and whenever possible, make sure they pay an up-front fee or deposit on the work to ensure you are not left months later having worked for free, still waiting on their first payment.

It is a very good idea to create a benchmark payment plan with your client if possible. If they have to pay when proof has been provided that certain stages of the project are completed, you are less likely to find yourself waiting months after the job is done with 50-100% of the bill still not paid to you.

Large corporations can be just as difficult to receive payment from at times and even make take longer to pay, so beware of being naïve with the impression that their billions of dollars in annual profit mean they will be faster, more reliable payers. It does not.

Conclusion

Technical communication may be performed in a wide variety of settings such as in a research center, a business of any size, in the field, in a university, or in a hospital to name only a few examples. Depending on the type of education you receive, your previous academic writing may be very different from the way you must write as a technical communicator.

In the professional world, your reader may be a very busy person who is only reading what you've written specifically for the purpose of gleaning the information they need. Often they require the data you are providing as a technical writer in order to determine what actions to take and or what decisions to make.

When writing a technical document, you must first and foremost be acutely aware of the purpose of the technical document you are working on, e.g. the purposed goal.

You will still need to apply a lot of what you learned in school in terms of grammar, spelling, punctuation, sentence structure and so forth, but there may be additional rules to follow as well. Your old English teacher may not have been qualified to edit a technical document.

Most readers of technical documents are unlikely to read the material in its entirety. Remember you're your readers are not a captive audience. Readers of your technical documentation will ask themselves questions such as: Why should I read this? And: How does it affect me?

Consider whether becoming a technical writer or technical communicator is really appropriate for your professional interests. However, if you work as a technician or engineer of some sort, you'll likely have no choice but to engage in some form of technical communication.

Dedication, attention to detail, the ability to casually

accept criticism, the willingness to continuously make improvements, and the acceptance of a generally moderate income level are just a few of the points to meditate on when determining if becoming a technical communications expert is a goal you would like to pursue.

Enjoying writing does not always mean you would enjoy spending 40 hours per week writing. Consider not only how much you actually enjoy writing, but also what the reasons are that you like to write. This can aid you in evaluating whether or not technical writing is the type of writing you would like to do and whether or not you actually want to pursue any kind of career in writing.

It can take years of practice to develop a comfort level with the work and facets involved in the process of technical writing. You should be willing to shut out the rest of the world for a few hours at a time to absorb yourself in the craft of sorting through words, concepts, and technical details associated with the document you will need to deliver.

Learning the mechanics and the concepts involved in technical writing may be an evolutionary process and like most people you'll find you become more comfortable with one aspect of technical writing than with a different aspect of it.

Of course, you may only need to develop a moderate comfort level as technical writing could easily be just one small facet of the work you perform and it might also just be something you only have to do once a year or less often depending on your career/work role. Hopefully you have found some of the information in this book useful to your interest and efforts in delivering a quality technical document for whatever purpose.

If you are seeking full-time work as a technical writer either independently or for an employer, the benefits of having the ability to work remotely from home, to build your own business, and be your own boss are certainly appealing to many. However, these definitely require self-discipline and dedication to achieve. But do not be fooled by articles talking about how you can make a great living as a technical writer. This is absolutely a matter of perception.

Yes, you can easily make more money as technical writer than you would if you worked at a fast food restaurant, as a restaurant server, as an accounts payable clerk, a receptionist, a bank teller, or any number of other general jobs available. But it is not a job that is going to make you wealthy without budgeting and lucky investments.

You may easily still find yourself in line with everyone else to purchase your lottery ticket hoping for some dream of financial independence and tired of the monotony of boring task work employment if you are working as a technical writer.

On the flip side of the coin, there are many of us out there who absolutely have to write. We are driven to write, good or bad, whatever the case may be, and as such, learning how to work as a technical writer and develop a successful career in this field makes sense.

You can make more money working as a technical writer than you can with some other forms of writing and you can secure full-time employment as a technical writer and you do have a lot of remote work opportunities available, particularly when you have developed a reputation as a reliable, skilled technical writer in one industry, or in a variety of industries.

As you begin your career as a technical writer, be very careful to remember the points I brought out in this book about CYA and organization and paying attention to standards and conventions within each industry. These are the keys to developing your reputation and continuing to work long-term in the field of technical communications.

Even if you are a technician or engineer of some kind where writing is just a regular part of your work and an extra task you hadn't thought much about when considering what your career would encompass, it is still imperative for you to maintain your reputation as a reliable, quality writer.

Continue to study various resources and books on the topic of technical writing so you may evolve your skillset and keep your skills up-to-date.

Take what you have learned in this book and build upon that knowledge as technical writers and technical communicators who grow their strength as writers and reliably turn out quality documents are huge assets to industry and have the opportunity to enhance the learning and daily work experience for countless others who rely on the documentation you will generate.

Glossary

Figure 12 – Artist: Chagouris - Copyright 2014

Abstract: An abstract may be called a summary. Key points from a document may be summarized in an abstract. An abstract may be 1-3 pages in length. An abstract may be an informational abstract, or it may be a descriptive abstract. There are different rules for each.

AP Style: AP stands for Associated Press and is the grammar and style guideline book for journalists such as newspaper writers and press release writers. While some rules of AP style overlap and match those of other style guidelines, following AP style is not correct for most technical writing work.

Applications: In relation to computing, applications are computer programs/software, generally considered to include a user interface.

B2B: Commonly used acronym for Business to Business sales. Marketing tactics in B2B sales differ in some respects from those used in B2C (Business to Consumer) sales. In B2B the customer of the business is another business, rather than the general public.

Bibliography: A bibliography may be an independent publication or it may be an appendix included in a book that provides supporting information associated with the subject or author. A bibliography is a list of resources/publications associated with a specified topic or author.

Blather: Blather is talking or writing with a lack of substance. It is the opposite of précis.

Branding Manager: Also called a Brand Manager – An individual assigned the role of policing corporate brand policy. This may include managing internal and external communications within an organization to ensure the brand message is consistent and clear. The brand manager will often help to prevent any images or other messages in various forms from being published in association with the organization that would potentially harm the company's reputation or that would contradict the focus chosen as the overall image the company should portray.

Case Study: A case study may include analysis and research. It may detail a process involved in developing a conclusion based on study of a particular case or a set of cases. A case study should be developed as the result of a study conducted over a period of time to evaluate a theory, behaviors, geological findings, any number of events, or other related matters associated with discovering information useful to a researcher and others. A case study is also considered to be an in-depth investigation.

Correspondence: Communication generally considered as communication between two or more people and in written form (either through the use of paper or electronic media).

CMS: Acronym for content management system – In this book used to describe what are also sometimes blogging systems such as WordPress, Drupal, Joomla, MODX, SharePoint, and DNN (Dot Net Nuke), to list only a few easy to install website applications where anyone willing to spend a minimal amount of time can create a website without knowing how to code in HTML or other languages, and from there can simply begin publishing web pages from a WYSIWYG environment similar to Microsoft Word.

Computing Industry: Businesses associated with hardware and software design, development, and distribution.

Conventions: A way in which something is supposed to be handled or managed; or a way an industry traditionally does something.

Creative Writing: Writing as an art form commonly associated with literature, poetry, screenplays, and varied fiction genres.

CYA: An acronym commonly used to denote the importance of taking actions for protection from the potentially negative results of misunderstandings, from others who may have malicious intent, or from computer malfunctions or other issues that can arise. It means retaining proof of communications and work.

Data Sheet: A data sheet is a thoroughly detailed explanation of technical characteristics of a specified technology, component, material, product, or property.

Editing: Editing is a detailed review including, but not limited to: document flow, content development components, consistency, clarity, and so on.

Ethnographic Record: Notes you may take in a methodic manner or via system of formatting and organization or other systematic means associated with cultural studies.

Expositing: Expositing is to expound or demonstrate reasons or explanations, perhaps to add supposed credibility in the perception of public eye.

Flowery Writing/Flowery Style/Flowery Writing Style: A distinctive manner of expression associated with using unnecessary words and language types such as may be found in fiction literature; Overly descriptive prose unnecessary to meeting the objective of a technical document and often a distraction to the reader attempting to gather information.

Grammar: Grammar is a set of rules defining acceptable language structure, syntax, formation of sentences via word/component part combinations.

Instructional Writing/Instructional Communication: Communication with the goal of teaching and/or informing.

Manufacturing Industry: Trade-based industry may include equipment, food products, machinery, textiles, and other facets of trade.

Non-fiction Writing: Written form of communication by an author or set of authors who consider the text factual and accurate.

Oil & Gas Industry/Energy Industry: Businesses involved in the distribution, generation, refining, extraction, production, and /or manufacturing of any form of energy or power.

Periodicals: Periodicals are publications that have been published during regular intervals such as daily, weekly, monthly, quarterly, or annually.

Précis: A much shortened version, an abstract, or a summary is a very basic definition of précis. The concept behind precise is eliminating the distracting blather and 'getting to the point'.

Preposition: A preposition connects a noun or a pronoun to other words – creates a connection between words. It is one of the parts of speech. A well-known saying is to never end a sentence with a preposition. However, this is a grammatical myth as there are times when a sentence technical should end with a preposition. A preposition should not be included at the end of a sentence when it is unnecessary. Additionally, many define a preposition as a word that should or could precede a noun. Obviously, if the preposition is at the end of the sentence, it is not preceding anything. A list of prepositions can be found here: https://www.englishclub.com/vocabulary/prepositions-list.htm

Process Engineer/Process Engineering: Concerned with industrial and production processes; May include process design, process production, process optimization, and process control; A liquid or gas is also called process and work of a process engineer may revolve around the transfer, transformation of, and other associated actions and affects to a process, such as the extraction of a process and such related.

Proofreading: Error discovery and correction; Detection of errors in grammar, spelling, and punctuation; Review of consistency in industry-specific terminology used and language sets (mixing of different forms of language – for example: authorise vs. authorize).

Qualitative: Measuring quality not quantity is one way to define the word qualitative. Observational research may be considered a form of qualitative research. Qualitative data may characterize, but not specifically measure.

Quantitative: Quantitative is related to measuring, as it is connected to quantity, rather than quality. This does not, however, mean that quality cannot be determined or estimated through analysis associated with quantity.

Repository: Centralized storage location generally but not always considered accessible to a group of people who are either members or the general public.

Resource: Source or supply of assets of some kind.

Software Industry: Computing industry specific to software design, development, and distribution.

STC: Society for Technical Communication (www.stc.org) – STC is a resource network for Technical Communicators.

Standards: In technical communication standards may refer to organizational standardization in the sense that specific language is required by use within an industry or organization for the sake of clarity in business. Standards govern that which is acceptable and not acceptable in terms of language, format, and various aspects of technical communication.

Tagging System: This is a system for cataloging equipment in the field that often includes a physical mark on a tool or piece of equipment and is noted in an electronic spreadsheet or database.

Technical Communication: Communication involving technical information and/or specialized information

Technical Documents/Documentation:
Specifications, user manuals and/or instruction sets, reference materials, informative articles and journals based on specialized data, industry/trade documents, various forms of reference material.

Technical Writing: Written form of technical communication.

Terminology: Words, phrases, terms, or expressions associated with specific theory, trade, study, professions, and industries.

White Paper: Not all white papers are technical; however they should be informative and are common marketing tools used in B2B. Technical white papers generally provide technical information about a product, technology, or service. While a white paper is expected to have more of a marketing slant to it, it should not be without an informative component. A white paper that provides no real information is an abuse of the technical consumer, or engineer's time to read.

Index

(Caveat: Due to the way the Kindle E-Book engine re-formats documents it is likely the index page numbers will not match the actual locations of the words/topics in the Kindle version. Please purchase the print version of this book in order to have a copy with accurate index indicators.)

Special Thanks To: My beautiful children who have grown up to become amazing individuals and who have been so supportive through these years as I've been working to grow as a mother, writer, artist, software engineer, and a myriad of other things it seems at times. Thank you to all of the contributors of words and art to this book, to those who read it, provided supporting words, tolerated my many requests to look at it as it has been developed, and to those who have tolerated my general excitement over topics only some of find exciting.

Also thank you to all of those many throughout these past 20 years who have hired me to write for them and in doing so provided me with so much experience to draw upon throughout the drafting of this instructional book to try and help others learn much more quickly what it apparently took me 20 years to learn. Also thanks to my mother and father whose life and death motivated me in so many ways to keep writing.

16090189R00134

Printed in Great Britain
by Amazon